GREEN GAMBIT

Climate change, climate policy, and the race against time

Victor R. Kalimugogo
and
Hassan M. Alzain

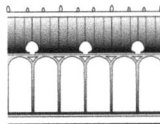

LONGUEVILLE
MEDIA

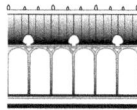

LONGUEVILLE
MEDIA

First published 2025
by
Longueville Media Pty Ltd
PO Box 205 Haberfield NSW 2045 Australia
www.longmedia.com.au

Print edition ISBN: 978-1-7640030-0-1
eBook ISBN: 978-1-7640030-1-8

About the Authors

Victor R. Kalimugogo
Sustainability Adviser
Victor is a passionate advocate for environmental management and has spent over 20 years delivering professional services in sustainability and risk management, primarily across Australia's eastern coast. His expertise spans advisory, compliance, and assurance roles across diverse sectors, including manufacturing, energy, resources, and infrastructure. He has worked with both public and private sector clients in Australia, New Zealand, South Africa, Germany, Kenya, Uganda, Tanzania, PNG, and Indonesia. He combines deep industry knowledge with consulting experience. Through this book, he hopes to inspire readers to unlock their sustainability potential.

Hassan M. Alzain
Environmentalist and Climate Advocate
Hassan Alzain is a dedicated champion of climate action and sustainability, combining his Yale University education with an unwavering commitment to tackling global environmental challenges. Actively participating in international climate discussions, he believes in the transformative potential of informed decision-making, innovative solutions, and collective effort to achieve a sustainable future. With an expanding collection of influential publications on climate science and conservation, Hassan seeks to inspire and equip readers with the knowledge and tools to drive meaningful change. Through this book, he aims to empower future leaders, policymakers, and communities to embrace sustainable practices for a better world.

CONTENTS

Preface

For many years, environmental issues, particularly climate change, have been leading issues in public and political discourse. Our understanding of the issues has been accompanied by the persistence of key positions of parties in the international arena. It is clear that climate change is strongly linked to socio-economic and cultural factors, and these relationships play out within the context of international climate negotiations as countries continually contend with each other for the most advantageous positions. Climate change has significant implications for all facets of our lives, and it is this multifaceted nature that presents difficulties to policymakers and leaders in designing universally beneficial policy instruments.

This book illustrates the dynamics of climate policy negotiations and the nature of the core issues. It does this by taking the reader through a series of building blocks, each of which examines a fundamental climate change issue discussed at the international level. The building blocks lay a foundation for the reader to more fully understand the dynamics of the nations in discussion.

The first building block is an introduction to the need for energy and a discussion of energy systems. The next building block is a look at the Industrial Revolution as arguably the greatest leap in human history and its effect on energy systems; the next building block is a look at climate change and what is meant by the term, including the impacts of climate change. If the previous building block can be considered as the scientific realm of climate change, the next block is the ideological aspect of climate change, such as environmental principles and climate negotiation processes. After considering the scientific and ideological aspects of climate change, it is important to consider how humans have viewed emerging sustainability issues and it is this issue that is considered in the next building block along with stakeholder interests and models of managing sustainability issues.

A key driver for us to write this book is our belief that solutions to climate change are deeply rooted in the idea of partnering. Industries are to collaborate with one another, and professionals are to find solutions through multidisciplinary working methods. Climate change needs to be viewed through a variety of lenses if the solutions are to have any meaningful impacts.

With the building blocks set, we then step up to look at the policy environment. There we see the fragmentation of multilateral climate negotiations into a growing number of blocs engaging each other via the United Nations Framework Convention on Climate Change (UNFCCC) mechanisms. Climate policy has always been a 'hot' topic of interest in the international arena, but in recent years it has become a highly contested area that has stirred the politics of special interests. It is clear that solutions to climate change will need international cooperation.

For those who endure frustration rather than faith in the multilateral process, it is instructive to note that big wins are more likely to be achieved through a multilateral process rather than bilateral or unconventional means. In the vexing issue of 'loss and damage' at the Paris COP for example, blocs representing small island states and the least developed countries were able to mobilise and win over the larger Group of 77 bloc to ensure the inclusion of loss and damage in the final agreement. True progress will likely rely on some form of the existing multilateral process, perhaps with some refinements in procedural mandates.

We close the issue of policy by pointing out the inherent challenges associated with climate change as a global public good. Reaching collective action without a central authority to finance and provide the public good is extremely challenging.

This conundrum serves as a useful transition into considering what the future holds. The last chapter explores future views and begins with an overarching discussion on the perennial issue of the north–south divide. Global South countries are generally thought to be more susceptible to the challenges of climate change such as extreme weather events, sea-level rise, and desertification. The chapter notes some of the reasons, mainly the geographical factors such as proximity to coastlines, altitude, latitude, and resource availability and ponders ramifications such as impacts on agricultural practices and climate-induced migration.

As the chapter winds down, we cast an eye towards the next century and what sort of climate to expect in 2101 and beyond. We speculate on the impact of sea level rise and the likely necessity for adaptation strategies. This speculation notes that ecosystems around the world are likely to experience significant changes in response to changing climatic conditions; we also note that changing climatic conditions are likely to affect societal systems – with a corresponding economic burden. These will require the deployment of significant technological and policy solutions.

The technological and policy imperative is already finding a voice in contemporary times, and the issues being discussed in the international climate arena reflect this dual imperative. The chapter highlights some of these issues, such as the global stocktake for monitoring and assessing the collective progress of nations in their climate commitments, the funding mechanism to support countries suffering from climate-related loss and damage, and the need to transform global food systems in the wake of the expected changes to climatic conditions.

We conclude the book with final thoughts on the divergence between collective acknowledgement of climate change and the fragmentation of preferred solutions. Rhetoric is one thing. Action is quite another matter altogether. The final words of the book hark back to famous past echoes of a former Norwegian Prime Minister.

Introduction

Earth is a dynamic celestial body with active internal processes which shape the surface existence. Geological, biological, atmospheric, and oceanic processes have interacted over billions of years to create the natural environment that sustains life. Throughout the evolution of the earth, the dynamism of these earth systems has resulted in significant challenges for its inhabitants, including humanity over the years. However, humans have so far successfully navigated these challenges and proved to be resourceful adaptors who can engineer solutions to ensure their survival. This resourcefulness has resulted in numerous anthropogenic issues, including but not confined to deforestation, fresh water depletion, water and air pollution, food scarcity, and others.

Among the issues, possibly the most topical one is climate change. It is a topical issue because it offers discussion points in multiple areas. Ideologically, there are discussions about beliefs around climate change, its existence and whether it is the most important challenge of modern times. There are other ideological issues, such as whether extreme weather should be attributed to the influence of humans, and, if so, then to what extent? Philosophically, the discussion on climate change is a polarizing one, even within the scientific community. From a policy perspective, there are climate debates about the usefulness of consensus for making effective policy. One reason for the interest in climate change is the indiscriminate impact of environmental aspects, they are not confined to a certain region only and are in fact often regional and global in their scale. Again, from a policy perspective, how should the problem be framed and once framed, who should take ownership of the solutions?

It is becoming clearer that humans are having a significant impact on earth systems. Attention on human-induced climate impacts has grown in recent years. The atmosphere and the oceans have warmed appreciably, resulting in sea level rise. There has also been a decline in Arctic sea ice, and glaciers are also shrinking. Global climate data shows levels of carbon dioxide (CO_2) have increased significantly since the Industrial Revolution.

Increase in global temperatures is often mentioned as a key indicator of climate change. Findings of the World Meteorological Organisation (WMO) suggest that a 1.1°C to 1.5°C rise in the global temperature should be expected in the next few years (2022–2026) from the average temperature between 1950 and 1990. Even though projected temperature increases are regularly discussed, the actual impacts of such increases deserve more attention. What would higher temperatures result in?

The increase in average global temperature is already resulting in more frequent high-temperature extremes, and events like heatwaves have already become more frequent and intense. Higher temperatures will also have a likely effect on the distribution and abundance of numerous plant and animal species. Agricultural productivity yields will most certainly be affected, and increased evaporation rates will also increase the prevalence of severe droughts.

In 2022, Europe and Asia experienced a foretaste of possible weather events and their devastating consequences. Western Europe and the United Kingdom experienced record-breaking heatwaves, and the month of June 2022 was the second-warmest June ever recorded for Europe. Meanwhile, heavy rains fell on regions of South Asia, including United Arab Emirates (UAE) and Pakistan, resulting in destructive floods. The impacts of the floods in Pakistan left over half a million people without adequate access to shelter, while also destroying large tracts of agricultural lands. Pakistan is a particularly powerful example of climate change issues because it has accounted for negligible amounts of historic greenhouse gas emissions, but it is often recorded by official agencies (such as the Asian Development Bank) as one of the most climate vulnerable locations.

...

In the recent past, climate change research has provided us with enough data. We already know a lot about climate change; for instance, the UN's Intergovernmental Panel on Climate Change

(IPCC) Sixth Assessment Report on the state of our climate asserts that the past decade was likely to have been the hottest period over the last 125,000 years.

The leading topics of climate change discussions typically involve environmental and social impacts, scenario modelling, and the mitigation and adaptation measures that are available as solutions. While these 'technical' issues are the leading conversations, there are other 'non-technical' but no less important issues that do not receive as much attention. For instance, concepts of equity and social justice have probably not received similar levels of attention as the other issues.

This book aims to provide the reader with an understanding in relatively equal measure of both the technical issues and non-technical issues that define our contemporary understanding of climate change. It hopes to furnish the reader with insights into observable climate phenomena, the disciplines and management processes to address the effects of these phenomena, and the ideological issues of key stakeholders such as leading institutes of climate research.

While climate change itself is an issue, we further address the related subjects, such as energy sustainability and the impacts of climate change policy on energy security.

In a nutshell, the book should provide useful reading and a quick source of information for the mature technical and non-technical professional, aid the purposes of the budding humanitarian, and be of interest to the everyday lay person. While addressing a wide range of subjects with specialized effort, every effort has been made to keep the book simple, though not simplistic, to ensure its readability and comprehension for an extended range of readers.

I

Introduction to Energy Resources

"Don't worry about me. I'm a survivor."

Types of energy resources

Life cannot exist without energy. When one considers the astrophysical views of the Big Bang theory for example, the origins of our universe owe their existence to the eruption of a cosmic mass which created elementary particles of matter and converted matter into three geneses of energy (nuclear, gravitational, and chemical energies) while the universe changed form. Shortly after the formation of the universe, during incredibly hot temperatures a large amount of energy was stored in the nucleus of atoms (nuclear energy); as the universe cooled, matter congregated further and resulted in the formation of planets and other bodies, resulting in gravitational potential energy between these bodies. Finally, as life on Earth developed, the process of photosynthesis enabled plant life to store solar radiation as a form of chemical energy.

Energy sources belong broadly to either of two categories – renewable and/or non-renewable energy. Renewable energy owes its existence to the sustainable natural sources and includes solar, wind, tidal, biomass, geothermal, and hydropower energy. Non-renewable energy typically lies underneath the earth and cannot be regenerated in any form once consumed; examples include natural gas, petroleum, coal, hydrocarbon gas liquids, and nuclear energy.

The major differences between the two sources are:

1. Regeneration: Renewable energy, as its name suggests, can be reused once consumed, while non-renewable energy cannot be available for reuse once it is consumed; it is for one-time use only.
2. Greenhouse gases: Renewable energy generates fewer greenhouse gas emissions in operations than non-renewable energy. This does not mean that it does not have environmental impacts. It may be 'clean' as we have come to know the word, but it is not necessarily 'green' on many occasions (hydroelectric power often has noteworthy environmental impacts for example, while nuclear power does not create any carbon emissions during generation).
3. Exhaustibility: Renewable energy is virtually an inexhaustible resource, whereas non-renewable energy has more finite reserves.

Having overviewed the two categories of energy sources, let us look at the main energy types within each of these categories. Note that the items explored within each category are not meant as an exhaustive list; rather, they are intended to provide the reader with a picture of the main energy types that have been preferred by humankind in past and present times.

Types of renewable energy sources

SOLAR ENERGY

Solar energy refers to any form of energy directly created from the sun. It is created when the sun's energy is converted into electricity or used for heating purposes. The sun is arguably the greatest source of energy available to humans. On a human timescale it is considered inexhaustible – it has not run short of its supply to the planet, despite having produced to the needs of living organisms for millions of years. It is expected to deplete its energy in billions of years.

There are two main types of solar energy applications:

1. Electrical: Photovoltaic (PV) materials and devices convert sunlight into electricity. Various PV systems can be installed on rooftops or integrated into structural designs of varying scales.
2. Thermal: Solar thermal refers to the process of converting solar radiation into heat (thermal energy). Such thermal energy can be carried on various mediums (e.g., air or water) and used for various applications, such as space heating.

Sunlight provides the earth's surface with an abundant amount of energy that far exceeds the needs of animals, plants, and all other living creatures. Solar energy now commands hundreds of billions of dollars of investor attention and is also attracting growing research and development focus. For instance, researchers and companies are testing the use of PV systems on railway tracks with a view to installing solar PV panels on the global railway network system.

WIND ENERGY

Wind energy is technically a by-product of solar energy. Humans have used wind energy for thousands of years already; however, in modern times wind power refers to the transformation of the wind's kinetic energy into electricity through the use of wind turbines. It requires wind farms to generate energy in the form of electricity, which can further be supplied for domestic and industrial use.

Wind energy is generated through the use of wind turbines to convert wind currents into other forms of energy. Turbines utilize the kinetic energy from the passing air currents through the rotational movement of the turbine blades, which are attached to rotors which, in turn, spin to form electricity. Wind energy applications are typically aimed at electricity generation. There are two main forms of wind turbines, namely:

1. Horizontal-axis wind turbines (HAWTs), which usually feature two or three blades that face directly into the wind.
2. Vertical-axis wind turbines (VAWTs), which place the main rotor shaft transverse to the wind. In a VAWT design, the blades are omnidirectional, so there is no need for a pointing mechanism as they capture the wind, regardless of direction. However, they are generally considered less efficient than HAWTs.

HYDROELECTRIC ENERGY

Hydroelectric energy is a commercially developed renewable energy source based primarily on large dam reservoirs being used to control water flow in order to drive a turbine, thereby generating electricity.

Hydroelectric energy is derived from the power of water in motion, such as flowing water from rivers, streams, or waterfalls (run of river) being channelled through water turbines to generate electricity. The flowing water exerts a pressure on turbine blades, which in turn cause shafts to rotate. The rotating shafts power electrical generators, which then convert the kinetic energy from the shafts into electrical energy.

Hydroelectric is considered a more reliable source of energy when compared to wind or even solar. In fact, it is probably the most mature renewable technology available, as it offers benefits such as lower greenhouse gas emissions and lower operating costs. Importantly, it also provides a quick ramp rate, so it can be used as both base or peak energy. Hydroelectric energy also has the advantage of being able to store excessive energy for future use. Tidal energy is also a form of hydroelectric energy and relies on the natural rise and fall of tides from the gravitational interaction between the earth, sun, and the moon to drive turbine generators.

GEOTHERMAL ENERGY

Geothermal energy refers to the extraction and utilization of heat from the earth's subsurface to generate electricity. Electricity can be generated from temperatures as low as 100°C (212°F), although it is more commercially optimal to tap into higher temperatures for large-scale applications. The earth's core lies nearly 3,000 kilometers (1,864 miles) below the surface and is considered the hottest part of our planet. Temperatures in the core can exceed 5,000°C (about 9,000°F), and this heat is in perennial outward radiation. As it radiates outward, it heats various geological substances and structures, such as rocks, water, and gas.

Geothermal energy is more effectively generated around medium or high temperatures, and this tends to be located in close proximity to tectonically dynamic environments. Low-temperature geothermal energy is available from virtually anywhere on earth just below ground levels in a thermal band of about 150°C (302°F).

Geothermal energy for electricity generation relies on heat sources a few kilometers below the ground. A most common application is to use natural sources of underground steam whereby steam is piped to power plants which have turbines to generate the required electricity.

BIOMASS ENERGY

Biomass energy refers to energy from living or previously living things, most commonly plants, wood, and municipal solid waste. It is one of the earliest forms of energy humankind has used. It involves the conversion of plant material-based solid fuel into electricity. Though the process involves the burning of organic materials and the release of carbon dioxide (CO_2), it is still a cleaner source of energy production, since the CO_2 is eventually restored by the regeneration of restored vegetation during photosynthesis. Biomass energy sources are the only renewable energy sources that can be used in liquid form (as biofuels). Biofuels are liquid fuels which originate from chemical conversion processes of biomass. The two most common biofuels are ethanol and biodiesel.

Ethanol (C_2H_6O) is an alcohol made from various plant materials. It is typically blended with gasoline to increase octane and reduce carbon monoxide and various forms of harmful particulate pollution.

It is usually blended in fixed ratios, because its octane-raising capabilities vary with the proportion of the blend and the chemical properties of the specific gasoline it is mixed with. Its energy content is about two thirds the energy content of gasoline, so it is not unusual for vehicles using ethanol blends to travel fewer miles per quantity of fuel than 100% gasoline equivalents.

Biodiesel is manufactured from vegetable oils (such as canola, camelina, soy, coconut, palm, and sunflower), animal fats (including beef or sheep tallow, pork lard, and poultry fat), or recycled restaurant grease. It is considered safe and produces fewer air pollutants than 'regular' petroleum diesel.

Biodiesel fuel blends easily with petroleum diesel. These biodiesel blends are described by their percent content of biodiesel, so that, for example, B20 contains a mix of 20% biodiesel and 80% diesel. With careful management, biodiesel can be used in diesel engines in transport vehicles, off-road mobile vehicles and equipment, and also stationary equipment.

Types of non-renewable energy sources

Fossil fuels such as oil, natural gas, or coal are derived from millions-year-old decomposed animals and plants. They are composed of only two elements, hydrogen and carbon, and are typically deep below the surface of the earth. Aside from those fossil fuels, nuclear fuel mainly based on uranium fission, also comes under the non-renewable energy category.

NATURAL GAS

Natural gas is a highly flammable colorless hydrocarbon primarily comprising methane and ethane as its most common and essential elements. Methane as a compound contains one carbon atom and four hydrogen atoms (CH_4). Ethane contains two carbon atoms and six hydrogen atoms (C_2H_6). Natural gas also contains smaller amounts of natural gas liquids – known as NGLs and used as inputs for petrochemical plants, burned for heating and cooking, and also blended into automobile fuels – and nonhydrocarbon gases, such as carbon dioxide and water vapor. Natural gas is consumed for different purposes, ranging from power production and transport fuel to cooking and heating. However, its uses are not confined to power and fuel purposes, as natural gas proves effective in certain chemical products such as dyes and fertilizers too.

PETROLEUM

Petroleum (crude oil) is the product of millions of years of high pressure and high temperature exerted on organic material. Petroleum reservoirs are generally located beneath land surfaces or on the ocean floors. In the early and middle parts of the 20th century, large oil reserves became available for commercial use, and since then, crude oil has been the major primary source energy around the world. As a fluid, petroleum can travel through the earth during its formation. The feasible commercial extraction of petroleum needs two components: an oil pool (the reservoir where the oil collects) and an oil trap (a rock formation that can hold the oil pool in place). Crude oil, and other hydrocarbons, exist in liquid or gaseous form in underground pools or reservoirs within sedimentary rocks of the earth's crust. Crude oil is generally refined into various petroleum products, including gasoline, diesel, and heating oil. Crude oil is associated with air pollution when it is burned, as well as greenhouse gas emissions from its combustion, accounting for significant amounts of carbon emissions.

COAL

Coal is a carbon-rich solid material formed through the compaction of plant remains. Coal's primary use is that of fuel – in fact, it was this very form on which the Industrial Revolution was based and advanced in the first place. Coal is sometimes associated with hazardous mining techniques and unsafe combustion outcomes. However, it is still one of the largest sources of energy for electricity around the world. Coal comes from underground formations (coal seams/coal beds) which can achieve a thickness of over 20 meters and a length of nearly 1,500 kilometers.

The creation of coal came by way of plant matter trapped in swampy lands. Over millions of years the plants were progressively buried under succeeding layers of mud and vegetation and became peat bogs. These peat bogs further developed into coal under more heat and pressure. Coal is therefore categorized by the degree to which it has matured over the formation process. Generally speaking, the deeper the coal seam is, the greater its calorific value (energy potential), and therefore its commercial value.

The combustion of coal liberates harmful gases and particulates into the environment. It also releases sulphur dioxide and nitrogen oxides, which manifest their toxicity through acid rain and smog, and contribute to respiratory ailments.

Evolution of energy sources

As the world grapples with global disruptions in energy markets and the shift toward a decarbonized energy future, it is instructive to consider the most consistently pressing energy issue of the 20th and 21st centuries: energy security. While the attention of policy makers and organizations at large has been fixed on cleaner energy options during the second decade of the 2000s, the reality of the interconnected world we live in ('global village') is such that energy security outweighs energy cleanliness when push comes to shove, and it has returned as a key requirement for sovereign nations. Geopolitical developments such as the Russia–Ukraine conflict have served to remind us of this uncomfortable truth.

Energy security refers to the absence of disruption to energy access as well as the affordability of this energy. Energy security is in natural alignment with the tenets of sustainable development and greener energy. However, energy security is also an entirely separate concept, and when energy issues become critical, its pre-eminence over other energy considerations becomes immediately apparent. This was demonstrated in the European energy policy response to the Russia–Ukraine conflict. While Europe was already well advanced in planning to phase out coal power plants to cut greenhouse gas emissions, the realities of the Russia–Ukraine conflict prompted the continent to restart coal plants to shore its energy supplies.

Societies have always been fascinated by the idea of energy security: it has played a significant role in the concept of civilizations through the ages. As the world considers its energy future, more questions are being asked about what form this future will take and what single clean, secure, and cheap energy source will power homes, industry, and public infrastructure. Does such a 'magical' fuel source exist in this energy future? It does not. The reality of the energy future is that it will be powered by multiple sources of energy, both renewable and non-renewable. As energy demand grows, societies will ensure that all available forms of energy are actually consumed, including solar, wind, petroleum, and other sources of energy.

ENERGY AND HUMAN DEVELOPMENT

Access to reliable sources of energy is often considered to be one of the key factors for human development. The story of human development has been consistently accompanied by efforts to convert energy from one form to a more desirable form (e.g., from wood to heating or lighting energy). Primitive humans consumed very little energy until the discovery of fire. As animals were domesticated for their labor and food, energy consumption of humans rose significantly during this agrarian era, especially with the advent of machinery and energy related to wind, water, and even coal. The industrial era spawned the use of the steam engine, which ultimately lay the foundations for a gradual adoption of full-scale coal, gas and oil energy sources.

The industrial era was a watershed of human development. It also ushered a new way for human communities to define themselves, based on economic growth – and with it, increased energy use. The 'developed economies' focused on economic growth and accumulated material properties in the process.

These developed economies can trace their economic success to their ability to convert energy into useful forms. Human developments during this period of the industrial era have witnessed a surge in achievements which may never be equalled in scale. Thus, this ability to explore and utilize different energy sources has brought unparalleled economic growth.

While this economic growth in developed economies has ushered unrivalled human development in material terms, it has also given rise to calls for 'degrowth' in the paradigm of continuous economic growth, ecological healing, and economic-environmental justice-based redistributions. The human development impact of these calls would be significant in any scenario. Ajl (2021) envisages three possibilities for the manifestation of reorientating the 'progressive practice' of continuous economic growth:

> ... [O]ne, full renewable replacement of current energy use, alongside increased southern energy use, and ongoing capitalist property structures – the left-liberal solution. Two, lower energy use through retrofitting core countries' infrastructures, substantial domestic redistribution to go back to 1950s-era levels of (in)equality, alongside full replacement of the energy infrastructure, and an ambiguous call for grants to help southern countries transition – the green social democratic and being-nice-to-the-South solution. And three, considerably lower energy use in the core alongside decommodified social infrastructure, guaranteed well-being, and massive technology grants to the Third World.

Moreover, since human development cannot be measured on economic growth alone, there are other, often underrated factors that must be counted. The most obvious are the environmental costs that need to be considered, such as various forms of pollution. Another factor is the extent to which technical advances can enable us to produce more with less. A key element of this discussion is the assumption that economic activity will increasingly rely on services, and that growth will therefore rely less on materials and energy. Evidence of current lifestyles, however, suggests that we will still be consuming significant materials and energy in the years to come. Thus, what we must understand is that energy sources have the polarity of both beneficial and adverse effects. Finally, there is the influence of societal cultures, whereby cultures in developed economies increasingly favor ecological attention over economic growth, while communities in developing countries often strive for economic material gains, often at incalculable environmental costs.

Despite great variations in access to energy and corresponding human development, it is clear that our overall access to energy today has undoubtedly been one of the greatest leaps in human development. Despite this great social leap, one third of the global population is still stuck in the wood economy – that is, it lacks access to the latest forms of energy. It is clear that the energy needs of this mass will need to be addressed by policymakers around the world as the shift toward decarbonization is progressively considered by policymakers and leaders around the world.

A shift to a more equitable energy landscape ('just transition') will necessitate significant socio-political-economic changes. The infrastructural aspects of our economy, like our transport systems and technological developments, are underpinned by social and political systems. Therefore, meaningful change will have to involve a change in social habits, which in turn will influence energy practices. It is poignant to consider the impact of social habits (and therefore energy practices) across the world. A combination of the BP Statistical Review of World Energy and the international energy data from the U.S. Energy Information Administration (EIA) shows the disparity in energy habits globally, as shown by the following examples (https://ourworldindata. org/grapher/per-capita-energy-use?time=2021):

PER CAPITA ENERGY CONSUMPTION
(KILOWATT-HOURS PER CAPITA)

Country	1965 (kWh)	2021 (kWh)
Algeria	2,006	15,496
Argentina	14,896	21,057
Australia	37,481	61,297
Canada	70,550	101,459
China	2,122	30,711
Colombia	4,864	10,372
Egypt	2,975	9,646
Ethiopia	253 (for 1980)	927 (for 2019)
France	27,189	40,489
Gabon	11,127 (for 1980)	7,603 (for 2019)
Germany	39,050	42,101
Iran	3,944	38,513
Japan	18,333	39,545
Kuwait	154,173	113,650
Malaysia	3,639	34,674
Netherlands	33,722	55,146
Norway	57,855	105,148
Oman	164	92,128
Rwanda	202 (for 1980)	491 (for 2019)
Senegal	1,796 (for 1980)	2,753 (for 2019)
Singapore	25,737	161,620
United Kingdom	42,711	29,641
United States	76,117	77,024

It is clear that requirements to change our energy habits in order to ameliorate ecological conditions mean different things to different parts of the world. The transition to lower-carbon economies will impact the developed economies in a different way from the developing economies. Ultimately, populations in countries with economic means will have to examine the extent to which they are willing to relinquish attachments of their modern living standards in the name of economic-environmental justice. Additionally, populations across the world have to examine the extent to which they aspire to economic accumulation models.

Currently, developing economies, including India and China, have committed themselves through different programs to providing low-carbon energy to as much of their respective rural populations as possible. Developed countries, having already ensured the provision of energy to their populations, can accelerate processes to help developing economies tackle their issues too. It is with this global mindset that we can minimize both the ratio of energy deprivation and, of course, its impact on our planet through global warming and climate change.

Meanwhile, a great mobilization of local resources, geo-political considerations, technological capabilities, and such other factors will be required to align developed and developing economies to ensure access to low carbon energy for populations, with a view to ensuring continued human development.

The concept of primary energy

Primary energy is the energy directly harvested from natural sources. It includes all those energy products which have not been transformed, directly exploited, or altered by any means. Fossil fuels (like crude oil, natural gas, and coal), biomass, solar radiation, solar mineral fuels, and hydroelectric, wind, and geothermic energies are the main sources of primary energy. In contrast, these primary sources can be converted to a secondary energy source, most often electricity. Transforming energy from one form into another, such as burning primary energy sources to generate heat in thermal power plants, is therefore not primary production.

All energy sources come from any of these primary sources, without exception. However, these sources, since they are in their basic form, further need to be converted to take a usable form for serving different purposes. For example, wind, coal, crude oil, and natural gas are all forms of primary energy from which electricity is generated/produced. So, electricity is the converted, and therefore usable, form of energy brought into its required form from primary energy sources.

Thus, all the human-derived forms of energy are primarily based on the transformation of natural sources of energy.

The BP Statistical Review of World Energy[1] is a global benchmark report that provides current information about the energy industry and global energy systems. It was first published in 1952 and is a useful source of information to understand the global primary energy mix. The 2022 publication covers energy data from 2021.

The report shows that fossil fuels accounted for 82% of primary energy, followed by hydroelectric power, renewables, and nuclear power. Oil still accounts for the majority of global energy sources in 2021, followed by coal, gas, and then hydroelectric power.

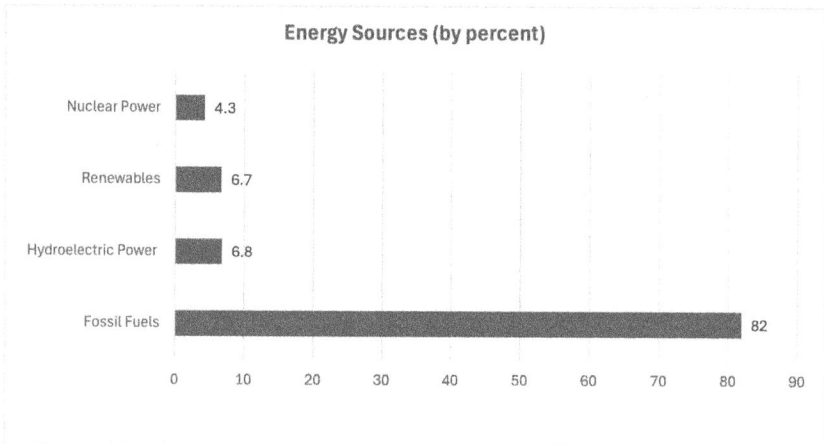

Energy Sources (by percent)

Source	Percent
Nuclear Power	4.3
Renewables	6.7
Hydroelectric Power	6.8
Fossil Fuels	82

Source: BP Statistical Review of World Energy

As primary global energy consumption grew by 5.5% from the previous year, there was a corresponding increase in global carbon dioxide emissions of 5.9% during the same period.

The consumption of primary energy is rapidly increasing, especially in the growing BRIC economies (Brazil, Russia, India, and China), whose development mainly relies on the combustion of primary energies. Economic growth and development clearly comes with a natural capital cost. Let us look at some of these primary energy sources.

1. BP Statistical Review of World Energy 2022, https://www.bp.com/content/dam/bp/business-sites/en/global/corporate/pdfs/energy-economics/statistical-review/bp-stats-review-2022-full-report.pdf.

Oil: In 2021, global consumption of oil stood at 94.1 million barrels per day (BPD). Oil consumption was just over 30% of primary energy consumption globally. Interestingly, refinery capacity declined for the first time in a generation, due mostly to a lack of adequate refining capacity to sufficiently process crude oil for the global market demand for fuels like diesel.

Natural gas: Consumption of natural gas globally grew by 5.3% in 2021. In fact, natural gas has experienced the fastest growth in demand for fossil fuels for a few years already. The US and Russia are the top two natural gas producers, globally.

Coal: Coal presents an interesting contrast in socio-economics. Because of the pollution hazards associated with its use, developed economies have shifted, or are in the process of shifting, from coal combustion. However, because of its cheap supply, developing economies still rely heavily on coal as a source of energy.

While global coal consumption has been on a downward trend for years already, due to the coal energy transition, it was still the dominant fuel for global power generation in 2021. Additionally, overall coal consumption in 2021 actually grew by over 6% from the prior year (noting that 2020 was the first year of the pandemic impact).

Renewable energy: Renewable energy is a term that refers to various energy sources known for their essentially limitless and/or 'green' supplies, including hydropower, solar, wind, geothermal, biomass, and wave and tidal energy. Renewable energy has experienced noticeable growth in recent years and, in 2021, global consumption grew by a significant 15%. Wind and solar, in particular, experienced the most material growth rates in consumption. In fact, for the first time ever, wind and solar energy combined to exceed 10% of global energy generation. Globally, hydropower is still the largest source of renewable energy and has powered many countries for decades.

Nuclear energy: Nuclear consumption in 2021 grew by 4.2% to its highest level in decades. It has been a key source of low carbon energy for many countries, and is the second largest source of low carbon energy for electricity, after hydropower. There are some significant differences in its usage patterns across various countries. For instance, in France it accounted for nearly 45% of Europe's total consumption, while in Italy that share was under 2%. Some countries, like Italy and Portugal, have negligible capacity of nuclear power reactors.

Nuclear energy is mainly used to generate electricity. In the US, 20% of electricity is produced from nuclear fission. Though the upfront capital costs are comparatively heavy, and the operation requires the installation of a series of nuclear reactors, the greenhouse gas emissions are very minor.

Nuclear reactors use pellets of uranium as their fuel, forcibly breaking atoms apart, resulting in the release of tiny particles, or fission products. These tiny particles contagiously help split other atoms of uranium, thereby leading to a chain of reactions. Heat is released and ultimately produces steam, forcing turbines to generate electricity by driving engines.

Biomass energy: In 2021, global biofuels consumption increased from the previous year by nearly 8%. This was mainly attributed to the growth of biogasoline consumption compared to biodiesel consumption over the period. However, for the 10-year period between 2011 and 2021, biogasoline consumption grew by just over 10%, while biodiesel consumption grew by just over 4%. By far, a great majority of the consumption and production of biofuels occurs in the two American continents, which accounted for well over 60% in both measures for 2021.

Biomass energy is generated from living (or once-living) beings. Plants are the most common sources of biomass energy. However, other sources include solid waste, landfill gas, and alcohol fuels.

Moreover, it has been the earliest source of energy used for heat in the form of woodburning. Today, it is used as a fuel for generators and machinery to produce electric energy. Biomass feedstock, including municipal solid waste and paper scraps, is also used for its thermal conversion to produce energy.

Biomass can also be processed into biofuel. Biomass is the sole renewable energy source that can be converted into liquid biofuels like ethanol and biodiesel. While ethanol is the result of biomass fermentation of high carbohydrates, including wheat, corn, and sugarcane, biodiesel is produced by combining animal fat with ethanol. Though biofuels, themselves, lack the power or capacity of gasoline, they can serve the purpose when blended with gasoline.

Biomass energy is considered to be carbon neutral, because it releases no more carbon when burned than it sequestered from the atmosphere. However, it should be noted that plants add significantly more value to the planet as living organisms than as fuels.

All these sources of energy, except for nuclear fuels, produce greenhouse gases which adversely affect the atmosphere of planet Earth.

Moreover, the consumption of primary energy is rapidly increasing, especially in the growing BRIC economies of Brazil, Russia, India, and China, whose development mainly relies on the combustion of primary energies. Thus, the growth and development come at a cost that merits concern.

Let us now discuss some of the key issues in transitioning from fossil fuels to renewables.

FOSSIL FUELS VS. RENEWABLE ENERGY: ARE WE STRUGGLING TO WEAN OURSELVES OFF FOSSIL FUELS?

We understand that the combustion of fossil fuels causes pollution, and the promise of renewable energies is to virtually eliminate this pollution. But the use of fossil fuels seems to be trending upward, even with the rapid growth of renewable energy. So, the question should be asked: Are we struggling to meaningfully cut our use of fossil fuels?

The impact of the pandemic is often cited to show how a brave new world could wean itself off fossil fuels. The year 2020 notably saw dramatic reductions in consumer activity and practically brought travel to a standstill. During this time, global greenhouse gas (GHG) emissions fell noticeably. In fact, according to the Global Energy Review 2020 from the International Energy Agency (IEA), emissions in 2020 fell by about 6%, which represents the greatest reduction since the Second World War. However, it is fallacious reasoning to suggest that crippling the global economy is a viable method of reducing GHG emissions and contributing to a better world. It simply fails the economic viability test and would adversely harm the livelihoods of too many people.

To understand why we are so dependent on fossil fuels, we need to understand how energy systems work. Fossil fuels were adopted en masse to replace mechanical and bio energy such as horse-drawn carriages. They formed from animals and plants which died prehistorically and were buried under rock layers and, in turn, became this source of energy after experiencing a prolonged decomposition process. As we have seen earlier, fossil energy comes in different forms, including coal, oil, and natural gas. Fossil fuels powered the Industrial Revolution and provided great swaths of populations around the world access to social amenities, through coal, oil, and natural gas, in that order.

But what attribute made fossil fuels a particularly attractive energy option for societies around the world, once their commercial exploitation was triggered? Their energy density and the convenience of moving them. Energy density is the amount of energy that can be liberated by a specific mass or volume of a fuel. The highest energy density fuel is hydrogen; however, since it is primarily a gas, it is difficult to store. Renewable energy sources have a lower energy density than fossil fuels, which means they liberate less energy for the same volume when compared to fossil fuels.

The energy density of various energy sources is shown below (adapted from C. Ronneau (2004), *Energie, pollution de l'air et developpement durable*, Louvain-la-Neuve: Presses Universitaires de Louvain):

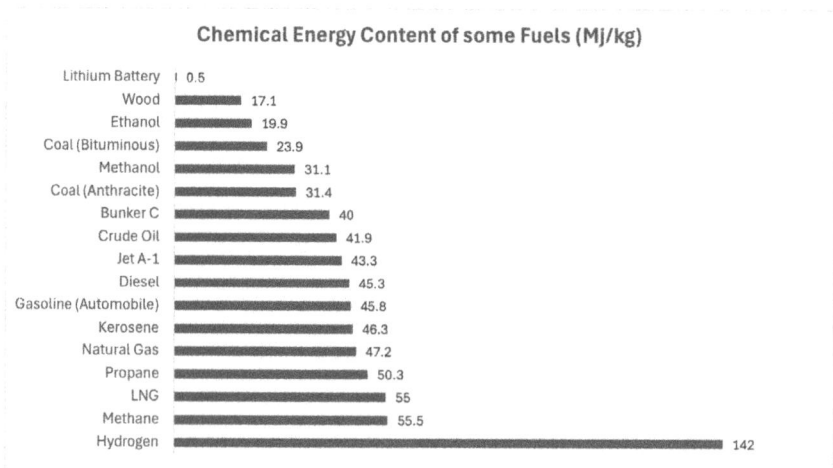

Chemical Energy Content of some Fuels (Mj/kg)

Lithium Battery	0.5
Wood	17.1
Ethanol	19.9
Coal (Bituminous)	23.9
Methanol	31.1
Coal (Anthracite)	31.4
Bunker C	40
Crude Oil	41.9
Jet A-1	43.3
Diesel	45.3
Gasoline (Automobile)	45.8
Kerosene	46.3
Natural Gas	47.2
Propane	50.3
LNG	55
Methane	55.5
Hydrogen	142

While fossil fuels pack a strong punch purely from the perspective of energy density, they still hold considerably less energy than nuclear fuels. However, nuclear sources have always been accompanied by significant health and safety concerns.

The convenience of fossil fuels has also been a significant factor. For instance, the liquid form factor of oil was a key enabler in the development of the internal combustion engine, and the transport and stationary energy sectors. The emergence of electricity power during the last few centuries proves the value of this convenience: while oil can be used for stationary electricity purposes, its liquid form allows it to be more easily transported and therefore more readily lends itself to use in the transport sector. However, coal is dense enough to be useful for stationary energy and is often used as a source of stationary electrical energy.

On the other end of the scale, while the economics of renewable energy are quickly stacking up, there are some realities that we still have to manage. Because of the relatively low energy densities of renewable energy, there are some serious questions that must be asked about land spaces that must be dedicated toward the production of this energy. Not only that, there are also environmental impacts associated with using this land; these impacts are often not insignificant, and in some cases, they are similar to the impacts of fossil fuel extraction.

We know from the work of various life-cycle analyses that renewable energy like wind, solar, and hydropower will require more raw materials, including iron, copper, aluminum, and cement per unit of energy produced than their fossil-powered electricity generation activities.

While renewable energy is not free of GHG emissions, the emissions are considerably less than those of fossil fuels. However, there are still questions about their ability to provide consistent power (intermittency of wind and sunshine, e.g.), and the ability of battery storage technologies to bolster peak energy demand, as needed.

Ultimately, the attraction of fossil fuels is their energy density and convenience. The energy history of human development through the ages has been one of humans seeking greater energy densities (e.g., from biomass to coal) and greater conveniences. To the extent that we are currently at the start of another energy transition, these factors are likely to hold true. Therefore, it is likely that humans will continue to use fossil fuels for as long as they provide the most optimum levels of energy density and convenience.

Though the transition from a conventional source that has been dominant for centuries to a totally new energy source may incur greater initial costs to begin with, the attraction of renewable energy is the promise of rapidly reducing costs and significantly less environmental impacts. This makes renewable energy particularly attractive, and many countries in Europe and America have already begun to transition.

However, failure to ensure justice considerations in other parts of the world may limit its global buy-in; this would be an unfortunate by-product of the global economy, and the environmental ramifications would also be a significant issue.

II

Industrial Revolution

"It's a mixed-use facility: retail space, low-rent housing, luxury apartments, and an area set aside for making steel."

If I had asked people what they wanted, they would
have said 'a faster horse!' – Henry Ford

The Industrial Revolution is arguably the single greatest leap human history has ever witnessed. Although it is termed a revolution, in fact the changes which occurred were gradual in nature. However, they were significant enough to fundamentally influence human endeavors in manufacturing, agriculture, transport, and social structures. What previously took months to accomplish through muscular energy suddenly took only days, and saved time and labor. These developments also fueled further advancements, as people learned more about industrial applications. Although the Industrial Revolution brought great economic growth and social opportunities, it also presented significant costs, including environmental damage and adverse exposure to health and safety hazards for workers. Just as significant as the scale of the changes was the rapidity of those changes. In a relatively short time, the Industrial Revolution generated a working class, established systems capable of faster mass production of goods, enabled greater access to these goods to a wide population, and spawned significant technological innovations.

As Austrian-American management guru and father of management thinking Peter Drucker states:

'Almost everybody today believes that nothing in economic history has ever moved as fast as, or had a greater impact than, the Information Revolution. But the Industrial Revolution moved at least as fast in the same time span, and had probably an equal impact if not a greater one.'

Brief background of the Industrial Revolution

The Industrial Revolution, which began in Britain in the late 18th century, had a significant impact on human life. It marked a significant shift from manual labor to machine-based production, leading to increased efficiency and productivity, and rapidly industrialized the leading economies of the world. This new way of producing goods and materials profoundly changed virtually every aspect of life.

The mechanization of a majority of manufacturing processes led to increased production and a reduction in prices. During the early to mid 1700s, the production of cotton textiles increased significantly, and prices fell drastically. While the mechanization of production processes for known goods was the early feature of the revolution, with time it created some completely new economic products and channels, such as the railroad, which truly revolutionized human mobility.

Military conflict was also impacted, with the production of weapons like cannons and muskets being rapidly increased and their costs falling significantly.

Socially, with the advent of factories, a working class was created. There were other socio-economic consequences, including the impact on slavery. The rapid growth of the textile industry (powered by the steam engine) had the effect of reviving the commercial imperative of slavery. The mechanization of the cotton industry created the need for low-cost labor and therefore increased the practice of breeding and dealing slaves. Another social impact was the increasing fragmentation of the family unit. In the artisan eras prior to the Revolution, it was not uncommon for family members to work together. The Industrial Revolution created a new workplace, with a new set of requirements, and that meant that some family members were left at home.

The environmental effects of the Industrial Revolution were especially significant. As industrialized mills and factories were set up for manufacturing, they needed larger amounts of water power. Traditional mills used dams, which were essentially temporary in nature and could be removed during certain times of the year, but the newer mills and factories relied on more permanent dams of a greater scale. These new dams had an immediate environmental impact, as they blocked migratory fish and flooded upstream meadows. For the first time, there were new and significant sources of air and water pollution.

The growth of factories and manufacturing plants meant that more pollutants were being emitted into the air, leading to air pollution. The burning of coal produced large amounts of sulfur dioxide and other pollutants, leading to the creation of acid rain, which is harmful to plants, animals, and humans. This pollution also caused respiratory ailments in people living near industrial regions.

The Industrial Revolution also brought about a significant change in land use. Large areas of forests were cleared to make way for factories and agriculture. This contributed to soil erosion, loss of biodiversity, and a decrease in overall air quality. These activities disrupted the natural balance of life, leading to a degradation of environmental systems, including soil, water, and air quality.

The impact of industrialization on water resources was also detrimental. The dumping of waste products into rivers and other waterways led to a decrease in water quality and affected aquatic life. Water scarcity was also a significant concern in many communities, mainly due to the over-extraction of water resources for irrigation and industrial use.

It is widely accepted that as the industrial and social network combusts these fossil fuel resources, they contribute to the rise in concentrations of greenhouse gases which may be overheating the planet. Scientists believe that prior to the Industrial Revolution, carbon dioxide levels were consistently around 280 ppm (parts per million) for almost 6,000 years of human civilization (National Oceanic and Atmospheric Administration, June 2022). Impacts on the oceans that are directly linked to the responsible human-induced industrial activities include increasing sea-surface temperatures, rising sea levels, and growing levels of acidification.

As the applications of energy from coal, petroleum, and natural gas increased over the decades, there was an increasing reliance on fossil fuel energy, one which still exists in contemporary times.

Drivers of the Industrial Revolution and links to climate change

Earth has a natural greenhouse effect due to the atmospheric presence of water vapor (H_2O), carbon dioxide (CO_2), methane (CH_4), and nitrous oxide (N_2O). These natural levels of greenhouse gases (GHGs) are essential for life on Earth as we know it. These gases allow solar radiation to reach the planet's surface and maintain temperatures at habitable levels. There is an important distinction between this natural greenhouse effect and the enhanced greenhouse effect (popularly known as climate change). The enhanced greenhouse effect (which we shall refer to as climate change from this point forward) is the additional radiative effect from increasing levels of greenhouse gases attributable to human activity. The gases whose levels are thought to be rising the most are CO_2, CH_4, N_2O, hydrochlorofluorocarbons (HCFCs), hydrofluorocarbons (HFCs), and even ozone in the lower atmosphere.

It is difficult to find consensus on the causes of the Industrial Revolution. Its effects have been profound. Indeed, work by Jon Steinsson ('How Did Growth Begin? The Industrial Revolution and Its Antecedents') identifies this impact very clearly:

'It is important, however, to realize that the era of rapid economic growth that we live in is a very recent phenomenon. Before 1750, economic growth was less than one tenth as rapid as it is today; and before 1500, economic growth proceeded at a truly glacial pace (as far as we can tell using current historical knowledge).

Our species has dominated the earth for thousands of years. Massive empires have risen and fallen. But over the millennia before 1500, the material well-being of ordinary workers changed very slowly if at all. Then, in a blink of an eye (from a long-term historical perspective), economic growth increased from close to zero to a modern rate of 2% per year. This dramatic change is what we call the Industrial Revolution.

The Industrial Revolution first occurred in Britain in the 18th and 19th centuries, but its roots extend back considerably in time and to other parts of the world. In this chapter, I seek to shed light on why the Industrial Revolution occurred, why it occurred in Britain, and why it occurred in the 18th and 19th centuries. Unfortunately, there is no consensus among scholars regarding these important questions. To the contrary, this is a topic of active, lively, and sometimes rather contentious debate....'

The impact of the Revolution is undeniable. By the middle of the 19th century, Britain (with a small fraction of 1% of global land area) contained under 2% of the global population and yet it produced two thirds of the world's output of coal and one half of the world's production of cotton textiles and iron (Clark, 2005). After starting in Britain, it swept through the rest of the western world and, progressively, its impacts were felt across the globe in one form or another. There was a surge in the production of manufactured goods, an explosion in population growth and attendant needs, significant increase in mass transport infrastructure, the creation of new social systems such as the factory and the labor class, the modernization of agriculture, and numerous other accompanying developments, such as mass urbanization. The world would never be the same again. Indeed, after experiencing significant changes in industry, demographics, agriculture, and transport, the world could never be the same again.

So, what is the link between the Industrial Revolution and climate change? Beyond the obvious fact of increased industrial and agricultural activity leading to increased emissions of GHGs, there were more nuanced factors at play. Firstly, the development of the steam engine had the effect of reducing production costs and growing profit margins in the industrial manufacturing sector. The coal industry was an immediate beneficiary of steam power and these increasing industrial developments. Coal mines had traditionally been limited by the requirement to pump out more groundwater the deeper the shafts became. Significant pumping power was needed to overcome this problem, especially as the coal industry had become a massive commercial enterprise. Consequently, investors poured money into the development of a solution to this problem. The major breakthrough came from a pump designed by Thomas Newcomen. By enabling coal mines to reach deeper into the ground, the Newcomen engine directly led to an expansion in the coal industry. The efficiency of this steam engine pump was improved by James Watt through a process that involved a 'double acting' engine which used high-pressure steam on both sides of the piston to double the output (and in the process, creating the term 'horsepower,' which is commonly used today). By the start of the 19th century therefore, there were thousands of steam engines in use across British mines, cotton mills, and industrial facilities. These engines used coal as an energy source, and that is the first credible link between the Industrial Revolution and climate change. Notably, these processes would be replicated in other western countries, thus demonstrating other aspects of climate change – its global profile and its scalability.

Secondly, the steam engine also stimulated the incredible developments of the transportation sector. The use of steam engines to power locomotives along railway tracks was a profound development, because it enabled the haulage of great quantities of goods and raw materials to various destinations at a much cheaper cost than the equivalent wagons or carriages. It is not an exaggeration to state that the steam-powered locomotive radically changed humankind's understanding of time, space, and productivity. The early steam locomotives were used to transport coal, but inevitably they were also used to transport passengers. Since steam that was used to power these new locomotives was generated by coal, the railway sector was a significant source of growth to the coal mining industry.

Similar developments took place in the marine transport sector where fossil-generated steam power would be used to drive ships. Up to the 1920s, coal was the primary energy source used to produce the steam to propel ships. Coal was eventually replaced by oil. These steam-powered vessels were used to transport raw materials and finished goods to their final markets in larger quantities and greater speeds. During the 19th century, steam ships ultimately led to great expansions in international trade and to vast migrations of people, especially to the United States.

These growing activity levels in industry and transport gave rise to the development of industrial cities (such as Manchester, in the UK) and an industrial workforce (a new form of working class).

Two causal factors of the Industrial Revolution and climate change have already been discussed, namely the developments in both industrial manufacturing and transportation, largely as a result of coal-powered steam engines. These activities entailed the increased combustion of fossil fuels. However, a third factor, the impact of consumerism as a force, cannot be ignored. Indeed, it can even be perceived as the ultimate source of climate change, once the wheels of industry began to roll. As Dr Matthew White points out, an increasing variety in clothes, food, and household items turned shopping into an important cultural activity in the 18th century. Improvements in transportation and manufacturing made 'opportunities for buying and selling ... faster and more efficient than ever before. And with the rapid growth of towns and cities, shopping became an important part of everyday life. Window shopping and the purchase of goods became a cultural activity in its own right, and many exclusive shops were opened in elegant urban districts: in the Strand and Piccadilly in London, for example, and in spa towns such as Bath and Harrogate' (White, 2014). The 18th century in particular witnessed an explosion in tastes for the consumption of certain goods. This explosion of tastes has even been described as a 'lust for objects' (Clark, 2010). The voracious appetites of this consumer lust in turn fed the need for growing industrial and transportation activity; as a result, consumer tastes can be said to be a very direct link to the climate change challenge of contemporary times. This is a profoundly important point, which has been somewhat overlooked as we search for modern solutions to what is not a modern problem at its core. Ultimately, it is the pervasiveness of this demand-side factor which must be at the heart of real solutions to the climate change problem. We shall revisit this at a later point.

EFFECTS OF INDUSTRIALIZATION ON THE ENVIRONMENT:

The Industrial Revolution has had far-reaching impacts on the environment, including:

1. Air pollution: The factories and industries that combust huge amounts of coal have led to high levels of air pollution, and resulting respiratory problems.
2. Water pollution: The dumping of industrial waste into rivers and other water bodies has led to polluted waters, and afflicted animal, plant, and human life with various diseases.
3. Land degradation: The clearing of vast areas of forests has led to soil erosion and an increase in the release of greenhouse gases like carbon dioxide and methane.
4. Climate change: Industrialization has contributed to the combustion of fossil fuels, which has resulted in trapped heat, causing symptoms of enhanced global warming (commonly referred to as climate change) such as rising global surface temperatures.
5. Habitat destruction: Industrialization has led to habitat destruction due to deforestation and rapid urbanization, causing thousands of plant and animal species to lose their habitats.
6. Loss of biodiversity: The loss of significant habitats has resulted in the extinction of numerous plant and animal species; such losses have led to a significant loss of our biodiversity capital.
7. Human health: Industrialization has directly or indirectly resulted in hazardous working conditions, leading to occupational hazards to human health.

To summarize, industrialization has led to significant economic growth and development, but also contributed adverse effects on the environment. We need to consider sustainable development models for long-term benefits and mitigate the impact of industrialization.

The Industrial Revolution roadmap in relation to climate change

Earth's climate has always fluctuated due to the natural forces that regulate our climate system. Recent changes in the earth's temperature, precipitation, sea levels, and sea ice indicate a cataclysmic shift in our environment.

> 'Climate change knows no borders. It will not stop before the Pacific Islands and the whole of the international community here has to shoulder a responsibility to bring about sustainable development.' – Angela Merkel

Average global temperatures during the Last Glacial Maximum (LGM) were approximately 12°C (54°F). During the subsequent interglacial period, the average global temperatures ascended gradually to 13.80°C (57°F). Since 1880, these averages have risen another 0.6 degrees to reach 14.4°C (58°F) as of 2015. This rate of warming is approximately 50 times quicker than the rate of warming over the last 21,000 years (Scotese, 2016).[2]

In 2021, the World Economic Forum, in collaboration with Visual Capitalist and climate data scientist Neil R. Kaye, published an article on the historical events that have accelerated climate change. The article is highlighted by a famous global temperature chart which illustrates the evolution of monthly global average temperatures over nearly 170 years. Temperature values have been compared to pre-industrial mean temperatures (1850–1900).

----

2. Scotese, C. R. (2016). 'Some thoughts on global climate change: The transition from icehouse to hothouse conditions' in the *Earth History: the evolution of the Earth System*. https://www.researchgate.net/publication/275277369_Some_Thoughts_on_Global_Climate_Change_The_Transition_for_Icehouse_to_Hothouse_Conditions (PALEOMAP Project, Evanston, IL., 2016).

Monthly global mean temperature 1851 to 2020 (compared to 1850-1900 averages)

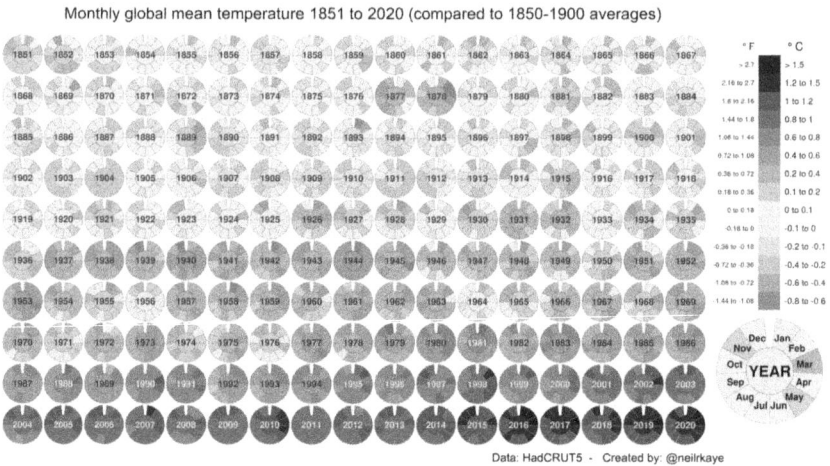

Data: HadCRUT5 - Created by: @neilrkaye

The IPCC, in its Summary for Policymakers as part of the Sixth Assessment Report, states that human activities, principally through emissions of GHGs, have unequivocally caused global warming, with global surface temperatures in the decade of 2011–2020 reaching 1.1°C above 1850–1900 levels.

While there have been small variations in this upward trending of temperatures, and there are also natural cycles that continuously exert their own forces on the climate system, the long-term trend since 1850 is quite clear and significantly more impactful than can be explained by purely natural phenomena. This is a rapid change in the context of Earth's climate system. Prior to these human impacts, natural changes in the climate system are thought to have happened over significantly longer time horizons.

As alluded to earlier in the book, the reason for the temperature increases can be demonstrated by the increase in atmospheric CO_2 concentrations which were consistently around 280 ppm for almost 6,000 years of human civilization. As human-induced emissions from various activities increased, the CO_2 concentrations also rose rapidly and in the 21st century these levels have gone above 400 ppm.

HOW DID WE GET HERE AND WHERE DO WE GO FROM HERE?

The manufacturing advances which kicked off the industrial revolution at the end of the 18th century revolutionized the way of modern life.

During the Industrial Revolution, the Global North increased its use of coal as a fuel source, thereby powering its economies through infrastructure such as new factories, transportation networks (e.g., shipping and rail), and manufacturing enterprises (e.g., iron smelting). Just as importantly, its usage increased the concentrations of CO2 in the air. Additionally, colonial ambitions also contributed to the release of GHGs through large scale deforestation in the name of commercial agricultural enterprises, urban growth and increasing industrialization. In the ensuing years "developing economies" have also embarked on their own development paths and the global profile of emissions has changed, particularly in recent decades. This is particularly true as a result of the larger economies such as China and India which benefited from the relocation of key production activities from the Global North to the Global South in the early part of the 21st century.

Consequently, the increase in CO_2 concentration levels has been accompanied by a corresponding increase in global surface temperatures, with fears of associated risks (e.g., risks to food supplies, natural habitat systems and mass migrations arising from extreme weather). It is this interplay of increasing CO_2 concentration levels, increasing surface temperatures and growing exposure to various risk factors which has attracted the attention of policymakers around the world.

Plotting the industrial revolution "roadmap" into the years ahead, it is extremely difficult to predict what future atmospheric CO_2 concentration levels might be due to the multiplicity of factors at play. Even if international climate policies and technological advancements enable the attainment of desired climate targets, Earth's own system of natural sources and sinks wields its own significant impacts. For instance, some plants grow more rapidly in a carbon-rich environments and the capacity of oceans to store carbon dioxide varies with their temperatures and circulation.

Given the uncertainty of future human activities (advances in policies and technologies) to understand the trajectory of this industrial revolution "roadmap" into the years ahead, one important issue remains to be considered: how rapidly would atmospheric GHG concentrations reduce if their emissions were reduced? The scientific reality is that the occurrence of atmospheric GHG concentrations as an outcome of reductions in their emissions depends on the chemical and physical processes that remove each gas from the atmosphere.

In the same Sixth Assessment Report, the IPCC reminds us that the concentrations of some greenhouse gases decrease almost immediately in response to emission reductions, while others may continue to increase for centuries even with reduced emissions. The point is given emphasis with the following words in the Summary for Policymakers: "Because of slow removal processes, atmospheric CO_2 will continue to increase in the long term even if its emission is substantially reduced from present levels. Methane (CH_4) is removed by chemical processes in the atmosphere, while nitrous oxide (N_2O) and some halocarbons are destroyed in the upper atmosphere by solar radiation. These processes each operate at different time scales ranging from years to millennia."

When we think of the history of global mean temperatures between the 1850s and the early part of the 21st century as visualized in the chart above, it is hard to argue against the impact of human activities since the industrial revolution. Looking forward, it is also clear that the key for the coming decades will be to find the most appropriate and responsible pathways to reduce atmospheric GHG concentrations as rapidly as possible.

Impact of World War II on industrialization

It is no surprise to learn that the Industrial Revolution had an impact on the conduct of war, and that war had an impact on the direction, scale, and undertaking of industrial and commercial activities. Equally, wars had an impact on the progress and direction of industrialization. The Industrial Revolution attracted and, progressively, even necessitated the use of the professional civilian in conflict. From scientific research and improvements to chemical applications to engineering capabilities, there was no limit to the importance of the civilian in wars, as wars became increasingly reliant on the advancement of technology and logistics.

At least for the early parts of the Second World War, the German advancements owed their successes not so much to new technological developments but, rather, to refinements in technologies and methods which had been used in the First World War. With the submission of France, the American industrial machine was compelled into action. A very significant and coordinated effort was kickstarted, involving scientists primarily from the US, the United Kingdom, and Canada. Their objective was to mobilize industrial resources to develop new technologies, boost research efforts, and develop retaliatory measures to enemy actions. The scale and success of this industrial mobilization was monumental and, in many ways, it was a precursor to modern-day multilateral cooperation.

It has been argued that wars can be significant sources of economic stimuli, and even reduce inequalities in regional development through the appropriate deployment of industries in regions of economic malaise. The US offers a strong example to support this theory of wars positively impacting economic activity and industrialization. World War II created a surge of economic activity so great that it essentially brought the US out of the Great Depression of the 1930s.

Significant developments of the allied industrial mobilization from World War II included radar communications, anti-submarine devices, rockets, and, of course, the atomic bomb. In the US alone, between 1940 and 1945, the war effort created an industrial economy so huge that it produced over 300,000 aircraft and bombers, 20,000 ships, nearly 90,000 tanks, and 350,000 trucks, as well as 9 million rifles and machine guns, and 40 billion bullets, all dedicated to aid 16 million soldiers (Klein, 2013).

The war mobilization efforts generated huge economic activity. Peacetime industries were turned to focus on war efforts. For example, automobile manufacturers made guns, trucks, tanks, and aircraft engines, instead of civilian cars. Shipyards built aircraft carriers, battleships, destroyers, and submarines. This sort of wartime economic activity stimulated the economy. Just as significantly, the technologies developed during the war had a very significant economic impact through the commercialization of new products which are still in use in modern times. These technologies included the following, to name a few:

Cavity magnetron: This device was developed initially by the British and, eventually, in collaboration with the US and Canada. It was a form of radar technology that enabled them to identify enemy vehicles (e.g., airplanes and ships) from a long distance under minimal visibility conditions. After the war, the very same device which was used to track enemy vehicles and improve Allied bombing accuracy also used electrical currents and magnetic fields to create electromagnetic waves in what we know as microwaves, the same waves used in microwave ovens. This radar technology was also used in the field of meteorology after the war. Using this radar, meteorologists have been able to develop their knowledge of weather patterns, thereby improving their weather forecast capabilities.

Computer technology: Although research and development into computers had begun prior to World War II, the war necessitated its accelerated development. One of the earliest computers developed for the war effort, but finished at the conclusion of the war, was the Electronic Numerical Integrator and Computer (ENIAC). ENIAC performed thousands of calculations per second and was released to consumers in 1946. This was the genesis of the computer as a mass tool of significant calculation powers, particularly in mathematics. The patent for this technology was eventually introduced into the public, and commercial development capabilities have since made significant gains in power, size, and application. In fact, 'Moore's law' became a term to define the observation of Gordon Moore in 1965 that the number of transistors in a dense integrated circuit (IC) doubles about every two years. Moore eventually co-founded Intel Corporation and this observation was the impetus of the semiconductor technology revolution that is still ongoing in modern days.

Mass medication: As the world has grappled with the effects of the COVID-19 pandemic, it is interesting to note how World War II spawned mass medication capabilities. Scientist Alexander Fleming is rightly credited with the discovery of penicillin in 1928. The antibacterial properties of this discovery significantly reduced fatalities from infections. However, it was not until World War II that the large-scale development of penicillin was commercialized by the US. The war made the mass production of penicillin a necessity, and scientists and researchers undertook numerous experiments, including with deep tank fermentation, to finally arrive at the method required for mass deployment of the medication. This served as a blueprint for future deployments of medication to the masses and was a significant milestone in the pharmaceutical industry.

It is clear that World War II had a significant impact on industrialization. Some wartime developments were so significant that they not only found application in peace but they impacted our interactions with technologies and our views of what was possible. World War II spawned new technologies, refined applications of existing technologies, and spurred massive growth in the reliance of research and development philosophies. So many parts of our daily lives as consumers trace their origins directly to World War II.

In conclusion, industrialization played a critical role in the outcome of World War II. The massive push toward increased production, innovation in technology, and expansion of transportation and infrastructure networks saw the major powers ramp up their industrial capabilities and helped to secure victory in favor of the Allied powers. Despite the destruction and devastation of war, the advances in industrialization from the war effort laid the foundations for continued advancements that would go on to change the world in ways that could never have been predicted.

Industrial Revolution and technology

The technological revolution and the sense of rapidly accelerating change began well before the 18th century and has continued to the present day. Perhaps the most distinctive aspect of the Industrial Revolution (IR) was its impact on enabling an increasing interplay between technology and industry. Significant advances in technology and industrial processes can be charted through the years; in fact, a series of industrial revolutions can be said to have occurred over the last few centuries. From the first Industrial Revolution (1IR) to the fourth (4IR), these developments have had the common outcome of leading to technical advances which have changed production and consumption methods.

The transition into production technology was entirely different than in the past. The new production technologies radically altered people's working conditions and way of life.

1IR: the first Industrial Revolution (understood by most people as *the* Industrial Revolution) essentially transformed agrarian societies into industrial powerhouses. From the middle of the 18th century onward, muscular energy was replaced by mechanized power as the use of the steam engine, coal power, and a growing knowledge of industrial processes provided humankind with an incredible leap in energy capabilities. These capabilities were characterized by the increasing use of machines to produce other machines. Key innovations in mechanical development led to transformations in a number of industries, primarily the cotton textile industry, energy generation through the increased exploitation of coal, energy usage through the steam engine (both for transportation and stationary energy purposes), the railway industry, and, finally, the iron and steel industry. Combined with changes in agricultural practices, new social structures characterized by a more influential wealthy middle class and an equally new accompanying working class, and increasing levels of urbanization, the effects of 1R were so profound as to effectively catapult these societies into unseen realms of material wealth and economic growth. Sustained population growth was an important element, if not direct cause, of 1IR. The significant demographic expansions taking place across the newly industrializing countries created ripe consumer markets which contributed to changing consumer patterns from household consumption to market consumption and beyond.

2IR: From the latter half of the 19th century, these newly industrialized countries experienced a wave of transformation through new sources of power and transportation which essentially built upon the foundations of 1IR. Additionally, there were significant developments in technological innovations and communications networks. In fact, this 2IR has often been labelled as the technological revolution. The emergence of new energy sources like electricity, oil, and gas also propelled the development of some of the most significant developments ever, including the internal combustion engine, and powered vehicles including automobiles and airplanes, and communications media such as the telegraph and the telephone. It was a key building block to the era of globalization of contemporary times. Transport and telecommunications infrastructure chipped away at limitations of time and distance, and in the process, they enabled much greater and more efficient modes of commercial activity while also revolutionizing the way people lived. The invention of the automobile captures the essence of these changing times. Automobile manufacturing was one of the earliest parts of the economy to use the assembly line, and after German inventor Karl Benz patented the world's first automobile in 1886, the concept of social movement had changed forever. The automobile gave people more capacity for leisurely freedom and greater access to economic opportunities. It also had a multiplier effect on other parts of the economy, such as the transport sector (through improvements and expansions to transport networks), automobile parts, and fuels (service stations). Increased leisurely freedom created the need for amenities such as hotels, restaurants, and fast-food chains to service new activities. Within a few decades, the automobile had entrenched itself as a key symbol of 2IR and transformed the world like virtually nothing else before it.

3IR and 4IR spectrum: The third and fourth ages of the Industrial Revolution are similar in the same way that 2IR built on 1IR, but are even more closely related.

The earlier phase of the 3IR–4IR spectrum was marked by the intersection of new communication and energy technologies; more specifically, it was driven by the complementary developments of processing, storage, and transmission of information with the harnessing of nuclear energy. It marked the beginning of the transition from mechanical or analogue electronic technology to digital electronics. During the earlier stages of its evolution, 3IR was marked by the increasing role of microprocessors and computing networks connecting with each other via software and hardware; computer-aided design, programmable logic controllers (PLCs), robotics, and telematics were among the inventions and developments that made a huge impact on all aspects of social and economic life. Later developments saw the emergence of nano technologies, 3D printing, artificial intelligence, and other forms of robotics. Perhaps most significantly, the internet announced itself to the world in growing and increasingly influential fashion. The internet has, itself, been a significant aspect of 3IR and has also enabled the capabilities of other innovations and industries.

The more modern-day aspect of 3IR–4IR is underway: the internet is joining with digitized energy infrastructure and digitized transportation and logistics networks and, in the process, creating an internet of things (IoT) platform. IoT is characterized by the use of sensors in modern devices, appliances, machinery, and other equipment. This network of physical 'things' connects and exchanges data with other devices and systems over the internet; the network and the digital systems within can record, monitor, and adjust each interaction between the connected things. Revolutionary developments in the development of computer processing efficiencies and power, combined with the prevalence of wireless networks in modern times, have resulted in the ability to turn all kinds of devices into parts of the IoT. This means that devices are now capable of interactions with each other and more responsive to various stimuli without the need for human interventions. IoT devices can range from items as innocuous as some children toys to more sophisticated equipment, such as connected manufacturing equipment. Because of the modular possibilities, it is possible to envision the possibilities for entire areas; indeed, the concept of smart cities will be based on this interconnectedness.

The 'Fourth' in the 4IR represents a fundamental shift in the manner in which we live, work, and interact. It is a new era in human development, made possible by technological advancements comparable to those of the first, second, and third industrial revolutions. These developments are fusing the physical, digital, and biological realms in ways that create both enormous promise and potential danger. This revolution is compelling us to rethink how countries develop, how organizations create value, and even what it means to be human, due to its speed, breadth, and depth. The true opportunity of the 'Fourth' in the 3IR–4IR spectrum resides in looking beyond technology and finding ways to provide the greatest number of individuals with the capacity to positively impact their families, organizations, and communities.

It is important to recognize the early achievements and possibilities: our ability to edit the building blocks of life has recently been vastly expanded by low-cost gene sequencing and techniques such as CRISPR (Clustered Regularly Interspaced Short Palindromic Repeats of genetic information); artificial intelligence is augmenting processes and skills in every industry; neurotechnology is making unprecedented strides in how we use and influence the brain as the last frontier of human biology; automation is disrupting centuries-old transport and manufacturing paradigms; and technologies such as blockchain, used in executive compensation, are reshaping the executive compensation industry. Global societal transformation is the consequence of these transformative developments.

Industry and economics experts have begun speculating about the Fifth Industrial Revolution, even as the Fourth Industrial Revolution is still in progress. The Fifth Industrial Revolution will entail saving the planet using the technologies of the Fourth Industrial Revolution 'for the good of the world,' with businesses appointing their own chief ethical and human use officers.

In addition, IT specialists predict that the Fifth Industrial Revolution, or Industry 5.0, will refocus attention on humanity. As a result, this industrial revolution will place a larger emphasis on human intelligence and the elimination of socio-environmental stagnation. In addition, experts anticipate that this period will be characterized by stronger connections between business and community purposes, with sustainability-minded brands gaining traction.

REFERENCES

Clark, Gregory. 'The British Industrial Revolution, 1760–1860,' ECN 110B, Spring 2005, World Economy History.

Clark, Gregory. 'The Consumer Revolution: Turning Point in Human History, or Statistical Artifact?' 2010, Working Paper, Munich Personal RePEc Archive (MPRA).

Klein, Maury. A Call to Arms: Mobilizing America for World War II. New York: Bloomsbury Press, 2013. https://scholar.google.com/scholar_lookup?title=A+Call+to+Arms%3A+Mobilizing+America+for+World+War+II&author=Klein+Maury&publication+year=2013.

White, Mathew. 'The Rise of Consumerism.' British Library, 15 May 2014, https://www.bl.uk/romantics-and-victorians/articles/the-rise-of-consumerism.

Ghosh, Iman. 'Since 1850, these historical events have accelerated climate change'. World Economic Forum. 9 February, 2021. https://www.weforum.org/stories/2021/02/global-warming-climate-change-historical-human-development-industrial-revolution

III

Climate Change

"I think it's time to consider the possibility that you might never reuse your old jars."

Early theories on greenhouse effect

Despite the scientific and policy controversy that the term 'climate change' attracts these days, the actual greenhouse effect is one of the more mature theories of atmospheric science. In fact, the broader understanding of the climate systems that impact Earth have been known (at least to a reasonable degree) much earlier than the IPCC First Assessment Report in 1990.

It was in the 19th century that scientists discovered that atmospheric gases caused a greenhouse effect, which warmed temperatures of the planet. Their interest at the time was the effect of changing carbon dioxide levels on ice ages of the earlier eras. These early scientists believed that a simple rock from the earth's distance to the sun should have a much colder temperature than the earth actually has.

EARLY ESTABLISHMENTS OF CLIMATE THEORY

In his 1827 *Mémoire sur les températures du globe terrestre et des espaces planétaires*, French physicist Joseph Fourier noted that energy, in the form of visible light from the sun, could easily penetrate the atmosphere and heat the earth's surface, which would then absorb some of this solar radiation. In absorbing this radiation, the earth's surface gains energy, some of which it loses through the emission of what we now know to be infrared radiation. The earth's atmosphere traps some of this infrared radiation and reflects some of it back to the ground. Fourier's work encouraged further research by other scientists in the realm of the type of influence that atmospheric gases might have on Earth's climate.

John Tyndall is credited as the next significant scientist to advance human knowledge in climate sciences. Tyndall was curious about whether some gases in the atmosphere could actually trap the radiation identified by Fourier. Through a series of experiments in 1859, he got confirmation that several gases could trap heat rays. The single most important one was water vapor. But the tests also confirmed the efficacy of carbon dioxide (CO_2) and methane (CH_4) in trapping heat.

But it was Swedish scientist Svante Arrhenius who was the first to establish a relationship between changing levels of CO_2 atmospheric content and ground temperature. Despite the acclaim in climate sciences, his actual education was in electrochemistry and he even received the Nobel Prize in Chemistry in 1903. Like other scientists of the time, Arrhenius was interested in understanding why the planet had experienced ice ages in earlier eras. While some scientists speculated and theorized on other reasons, such as astronomical factors, Arrhenius was of the opinion that atmospheric CO_2 could be a determinant of the glacial cycles. Through some research on available data and performing a series of calculations, in 1896 he established his theory: he proposed that if atmospheric CO_2 could be halved, it would lead to a reduction in average temperature. He was therefore the first person to propose a relationship between atmospheric CO_2 concentrations and temperature. He discovered that the natural greenhouse effect – the average surface temperature of the earth – was about 15°C as a result of the heat-trapping properties of water vapor and CO_2. He went on to suggest an enhancement of this effect by asserting that a doubling of the CO_2 concentrations would lead to a 5°C temperature rise.

SUBSEQUENT SKEPTICISM

This topic was put in the background for decades after. For one thing, the belief was strong that human influences could not compare to the scale of natural systems such as ocean circulation. Additionally, the oceans were considered to be carbon sinks of such scale that any human pollution could be 'offset' by the natural processes. In any case, it was assumed that water vapor was the chief greenhouse gas.

In the 1940s, this thinking was debunked with some remarkable findings. Canadian physicist Gilbert Plass predicted the increase in global atmospheric CO_2 levels and their impact on average surface temperatures. Developments in infrared spectroscopy for measuring long-wave radiation proved that increasing the amount of atmospheric CO_2 directly resulted in increased absorption of infrared radiation. Additionally, the research also confirmed that water vapor had a completely different absorption profile than CO_2. Plass therefore concluded that increasing levels of atmospheric CO_2 would intercept infrared radiation, thereby increasing the warming effect on the earth.

AGE OF CLIMATE INFORMATION

Plass was certain that if his theory was correct, the impact of global warming would start to manifest beyond natural processes at the start of the next century. Once meteorologists also exposed the fallacy of infinite CO_2 adaptive capacity of the oceans, the need for information to validate theories by scientists like Plass took on more urgency. Suddenly, it was openly acknowledged by researchers and scientists that the importance to more accurately measure the CO_2 atmospheric levels could not be overstated. Unfortunately, leading methods of the time were inadequate to provide any useful information, as they were, themselves, responsible for significant amounts of the 'noise' in results.

American scientist Charles David Keeling had other ideas, and decided to remove the sources of noise. Keeling secured funding for new equipment and initiated a series of samples at the isolated areas of Antarctica and high atop the Mauna Loa volcano, in Hawaii. Through these new locations he was able to establish clean baseline information on atmospheric CO_2 levels and plot CO_2 concentration curves.

In 1960, with two years of data after setting up in the new locations, Keeling reported that this baseline level had already risen. Not only that, the increase was at a rate that could reasonably be expected in the absence of oceanic absorption of industrial emissions. Climate information was now in the arena of not just scientists but also policymakers and the mainstream public. The Mauna Loa curve is still referenced by scientific journals today.

CO$_2$ IN THE MAINSTREAM

Ocean sediment research in the middle of the 20th century onward suggested that there had been multiples of cold–warm cycles in the last couple of million years. As a result, opposite fears of global cooling began to take hold. Additionally, different views on climate science became apparent. While some scientists were convinced about the effects of CO_2 on the climate, and would even make perhaps the first official claims of adverse climate change by the new century through President Lyndon B. Johnson's Science Advisory Committee, other scientists thought of CO_2 as merely one in a number of biological, oceanographic, or meteorological factors that could determine climate change.

A number of events occurred in the 1970s and 1980s which put the climate change theories back on the agenda. Wallace Broecker, a geologist by training, published a scientific article, 'Climatic Change: Are We on the Brink of a Pronounced Global Warming?' in 1975. His primary interest was always in the ocean's role in climate change and how an interruption of ocean circulation patterns could lead to climate change. Additionally, research in the 1970s and 1980s confirmed the significantly greater radiative warming effect of methane and other gases. Interest in the carbon cycle surged, and the work of Keeling continued to provide data for the work of scientists and, increasingly, policymakers.

In the 1980s, when global mean temperatures began to rise, the mainstream acceptance of the global cooling theory turned to skepticism. Activists took on the challenge further and began to advocate global environmental protection to prevent further human-induced climate change. The media also took an active interest in climate change, an interest that prevails to this day, possibly even in greater scale.

In the 1990s and 2000s, the era of computer models emerged as a force to determine climate patterns. Climate models, also known as General Circulation Models, or GCMs, are computer simulations of the earth's climate system, including the atmosphere, ocean, land, and ice. These models are based on well documented physical processes which simulate the transfer of energy and materials through the climate system. They calculate numerous properties of the climate system, such as temperature, pressure, wind, and humidity, for thousands of points on a three-dimensional grid.

Running these models is a technically complex process that involves representing variables of the climate system with mathematical equations and solving these equations with powerful computer technology. Once the outcomes of the equations can be tested to satisfaction, the results for future climate scenarios are then assumed to be useful to decisionmakers and used by policymaking bodies like the Intergovernmental Panel on Climate Change (IPCC). In fact, the IPCC suggests a number of criteria that should be met by climate scenarios if they are to be useful for impact researchers and policy makers (IPCC website):

Criterion 1: Consistency with global projections. They should be consistent with a broad range of global warming projections based on increased concentrations of greenhouse gases. This range is variously cited as 1.4°C to 5.8°C by 2100, or 1.5°C to 4.5°C for a doubling of atmospheric CO_2 concentration (otherwise known as the 'equilibrium climate sensitivity').

Criterion 2: Physical plausibility. They should be physically plausible; that is, they should not violate the basic laws of physics. Hence, changes in one region should be physically consistent with those in another region and globally. In addition, the combination of changes in different variables (which are often correlated with each other) should be physically consistent.

Criterion 3: Applicability in impact assessments. They should describe changes in a sufficient number of variables on a spatial and temporal scale that allows for impact assessment. For example, impact models may require input data on variables such as precipitation, solar radiation, temperature, humidity and windspeed at spatial scales ranging from global to site and at temporal scales ranging from annual means to daily or hourly values.

Criterion 4: Representative. They should be representative of the potential range of future regional climate change. Only in this way can a realistic range of possible impacts be estimated.

Criterion 5: Accessibility. They should be straightforward to obtain, interpret and apply for impact assessment. Many impact assessment projects include a separate scenario development component which specifically aims to address this last point.

As climate data matures into climate information and gains even further mainstream acceptance, there is an inconvenient truth that has emerged from the numbers. If the global community is to stabilize temperatures as per the aims of the Paris Agreement, the total of all carbon emitted from the Industrial Revolution into the future must not exceed the sum of a 'carbon budget.' It appears that over half of this budget has already been used up, primarily by industrialized economies but with growing proportion of that consumption by emerging economies.

Understanding climate change

Climate change is an intricate and pressing issue that demands a thorough understanding of its causes, consequences, and potential solutions. This section explores the multifaceted aspects of climate change, while also highlighting notable theories associated with this global phenomenon. By delving into the scientific exploration of climate change, we can gain insights into the challenges it presents and the steps needed to address it effectively.

Causes of climate change

Climate change has emerged as one of the most pressing challenges facing our planet today, with far-reaching implications for ecosystems, economies, and human well-being. The earth's climate is influenced by various natural and anthropogenic factors, resulting in alterations to global temperature patterns, precipitation levels, and weather events.

Climate change is predominantly driven by human activities that contribute to the increase in greenhouse gas concentrations in the atmosphere. The burning of fossil fuels, deforestation, industrial processes, and agricultural practices release significant amounts of greenhouse gases, such as carbon dioxide (CO_2), methane (CH_4), and nitrous oxide (N_2O). These gases trap heat in the earth's atmosphere, leading to the intensification of the greenhouse effect and subsequent global warming.

This following section delves into the multifaceted causes of climate change, shedding light on both natural phenomena and human activities that contribute to this global concern.

1. Natural causes of climate change: Climate change is not a new phenomenon, as the earth's climate has fluctuated throughout its history. Natural factors have played significant roles in shaping these variations, including:

a. Solar radiation: The sun's energy output undergoes periodic variations, which can impact the earth's climate. Changes in solar radiation intensity can influence temperature patterns over short and long timescales.

b. Volcanic activity: Volcanic eruptions release substantial amounts of gases and particles into the atmosphere, leading to temporary cooling due to increased reflection of sunlight. However, volcanic emissions also release greenhouse gases, contributing to long-term warming effects.

c. Orbital variations: Cyclic changes in the earth's orbit around the sun, such as eccentricity, axial tilt, and precession, influence the distribution of solar energy received by the planet. These variations are responsible for natural climate oscillations over thousands of years.

2. Anthropogenic causes of climate change: Human activities have become the dominant drivers of climate change in recent times. The following factors highlight the impact of human actions on the climate system:

a. Greenhouse gas emissions: The burning of fossil fuels for energy production, industrial processes, transportation, and deforestation has significantly increased the concentrations of greenhouse gases (GHGs) in the atmosphere, such as carbon dioxide (CO_2), methane (CH_4), and nitrous oxide (N_2O). These gases trap heat, resulting in the enhanced greenhouse effect and subsequent warming of the planet.

b. Deforestation: Forests act as carbon sinks, absorbing CO_2 from the atmosphere through photosynthesis, Widespread deforestation, particularly in tropical regions, reduces the earth's capacity to sequester carbon, contributing to increased atmospheric CO_2 levels.

c. Agriculture and livestock: Intensive agricultural practices, including rice cultivation and the use of synthetic fertilizers, release significant amounts of CH_4 and N_2O. Moreover, livestock farming, especially cattle rearing, produces CH_4 through enteric fermentation and manure management.

d. Industrial processes: Industrial activities release GHGs through the burning of fossil fuels, as well as emissions of potent synthetic gases like hydrofluorocarbons (HFCs) used in refrigeration and air conditioning systems. These gases have much higher warming potentials than CO_2.

e. Land use changes: The conversion of natural ecosystems, such as forests and wetlands, into urban areas or agricultural land alters the earth's surface and affects regional climate patterns. Changes in land cover also contribute to the loss of carbon sinks and the release of stored carbon.

f. Waste management: Improper waste disposal, particularly the decomposition of organic waste in landfills, generates CH_4 emissions. CH_4 capture and utilization techniques can help mitigate these emissions.

Climate change is a complex phenomenon driven by a combination of natural processes and human activities. While natural causes have historically influenced climate variations, anthropogenic factors have become the primary drivers of recent climate change. The increased concentration of greenhouse gases in the atmosphere, resulting from the burning of fossil fuels, deforestation, and intensive agricultural practices, is the most significant human-induced contributor to global warming in present times.

OCEAN ACIDIFICATION THEORY

Ocean acidification is a process resulting from the increased absorption of CO_2 by the world's oceans. As human activities release more CO_2 into the atmosphere, a portion of it is absorbed by the ocean, causing a chemical reaction that lowers the pH of seawater. The theory of ocean acidification explains how this phenomenon can have far-reaching consequences for marine ecosystems and organisms.

When CO_2 dissolves in seawater, it forms carbonic acid, which increases the concentration of hydrogen ions (H^+). These additional hydrogen ions lower the pH of the seawater, making it more acidic. The increased acidity affects the availability of carbonate ions, which are essential building blocks for marine organisms, such as coral reefs, shellfish, and other calcifying organisms.

The theory of ocean acidification suggests that the decreasing pH and reduced carbonate availability can impede the growth and development of marine organisms that rely on carbonate minerals to build their shells, skeletons, and protective structures. This can lead to weakened structures, increased vulnerability to predation, and diminished reproductive success. Coral reefs, for example, are particularly vulnerable to ocean acidification, as the reduced carbonate availability hinders the ability of corals to build and maintain their calcium carbonate skeletons.

Furthermore, ocean acidification can disrupt marine food webs and biodiversity. The changing chemistry of seawater can affect the physiology and behavior of various marine species, including fish, plankton, and other primary producers. This can have cascading effects throughout the ecosystem, impacting the abundance and distribution of species and altering the dynamics of marine communities.

Understanding the theory of ocean acidification is critical because it highlights the interconnectedness between climate change and the health of marine ecosystems. The reduction of CO_2 emissions and the implementation of sustainable practices are essential for mitigating the impacts of ocean acidification and protecting the delicate balance of marine life.

In conclusion, the theory of ocean acidification explains how increased CO_2 emissions lead to the absorption of carbon dioxide by the oceans, resulting in lower pH levels and reduced availability of carbonate ions. This phenomenon can have detrimental effects on marine organisms and ecosystems, highlighting the urgency of addressing climate change and implementing measures to reduce greenhouse gas emissions. By comprehending the theory of ocean acidification, we can better appreciate the far-reaching implications of climate change on the world's oceans and work toward sustainable solutions to mitigate its impacts.

Consequences of climate change

Climate change, a phenomenon primarily driven by human activities, has become a pressing global concern. This topic delves into the consequences of climate change, providing a detailed scientific analysis backed by empirical evidence, practical examples, and underlying theories. By examining the various facets of climate change impacts, this section aims to highlight the severity and urgency of addressing this global challenge.

RISING GLOBAL TEMPERATURES

One of the most prominent consequences of climate change is the steady increase in global temperatures. Scientific evidence, such as data from temperature records, satellite observations, and ice core samples, overwhelmingly supports this claim. The Intergovernmental Panel on Climate Change (IPCC) states that human activities, particularly the burning of fossil fuels and deforestation, have significantly contributed to the rise in greenhouse gas concentrations in the atmosphere.

The consequences of rising temperatures include:

a. Melting polar ice: The Arctic and Antarctic regions are experiencing accelerated ice melt. As the polar ice sheets diminish, sea levels rise, endangering coastal ecosystems and human settlements. According to the National Aeronautics and Space Administration (NASA), arctic sea ice has declined at an average rate of 13.2% per decade since 1979.

b. Ocean acidification: Increased atmospheric carbon dioxide (CO_2) concentrations lead to higher CO_2 absorption by the oceans, resulting in ocean acidification. This has profound implications for marine life, including the degradation of coral reefs and disruptions in the marine food chain. The Ocean Acidification International Coordination Centre reports that the pH of surface ocean waters has decreased by approximately 0.1 units since the beginning of the Industrial Revolution.

EXTREME WEATHER EVENTS

Climate change intensifies the frequency and severity of extreme weather events, including hurricanes, droughts, heatwaves, and heavy rainfall events. The scientific consensus indicates that a warmer climate enhances the likelihood of such events, as explained by the following theories:

a. Increased energy in the atmosphere: Warmer temperatures provide more energy to fuel extreme weather phenomena. For example, hurricanes thrive on warm ocean waters, which have been growing warmer due to climate change. The fifth assessment report of the IPCC concluded that there is a high probability of increased intensity and duration of hurricanes in a warmer climate.

b. Disrupted atmospheric circulation patterns: Climate change can alter atmospheric circulation patterns, leading to shifts in weather systems. This disruption may result in prolonged droughts, heatwaves, and/or unusual precipitation patterns. The disruption of the jet stream, for instance, has been linked to an increased frequency of prolonged heatwaves and cold spells. The attribution of specific events to climate change, however, remains an active area of research.

IMPACTS ON ECOSYSTEMS

Climate change poses significant risks to ecosystems, disrupting biodiversity and ecological balance. The following examples illustrate the consequences:

a. Species extinctions: Rapid climate change forces many species to adapt or migrate to more suitable habitats. However, the speed of climate change outpaces the ability of certain species to adapt, resulting in local extinctions and ecological imbalances. The International Union for

Conservation of Nature (IUCN) estimates that one fourth of the earth's species could face extinction by 2050 if global temperatures continue to rise.

b. Range shifts: Climate change alters the geographic ranges of various species. For instance, certain plant and animal species are moving toward higher altitudes or latitudes to escape unfavorable conditions. These range shifts can disrupt entire ecosystems and lead to the loss of crucial ecological interactions. The migration of species to new areas can also introduce competition, predation, and disease transmission, further destabilizing ecosystems.

AGRICULTURAL IMPACTS

Climate change affects agricultural productivity, jeopardizing global food security. The following are some of the consequences observed:

a. Decreased crop yields: Extreme weather events, such as droughts or floods, can damage crops and reduce yields. Rising temperatures and shifting rainfall patterns also impact agricultural productivity. The Food and Agriculture Organization (FAO) reports that climate change has already reduced yields of staple crops, such as wheat and maize, in many parts of the world.

b. Changes in pest dynamics: Climate change alters the geographic distribution and behavior of pests and diseases that affect crops. This increases the risk of crop losses and affects farmers' livelihoods. For instance, warming temperatures allow certain pests to expand their range into new areas, leading to the infestation of previously unaffected crops.

A significant theory associated with climate change is the concept of positive feedback loops. Positive feedback loops occur when a change in one component of the climate system amplifies the initial change, leading to further impacts. For example, as global temperatures rise, arctic ice melts, reducing the earth's albedo – the reflectivity of the surface. With less ice to reflect sunlight, more heat is absorbed, further accelerating warming. This positive feedback loop enhances the rate of climate change and highlights the interconnectedness of the earth's systems.

The consequences of climate change are far-reaching and demand urgent attention. The scientific evidence, examples, and theories presented in this section highlight the severity of the issue. Mitigating climate change requires collective efforts, including transitioning to renewable energy sources, implementing sustainable land use practices, and adopting climate-resilient policies. Additionally, raising awareness and promoting education about climate change impacts can inspire individuals, communities, and governments to take action. By understanding the consequences and acting upon them, we can strive for a sustainable future and safeguard the planet for generations to come.

Hazardous impacts that may lead to our extinction

GLOBAL TEMPERATURE RISE

a. Heatwaves: As global temperatures rise, heatwaves become more frequent, prolonged, and intense. This extreme heat poses a direct threat to human health and can lead to heat-related illnesses, heatstroke, and even death, particularly among vulnerable populations such as the elderly, infants, and those with pre-existing health conditions.

b. Crop failures: Higher temperatures can have detrimental effects on agriculture. Heat stress reduces crop yields and lowers the nutritional value of crops. Prolonged droughts and increased water evaporation can also lead to water scarcity for irrigation, resulting in crop failures and food shortages.

c. Disrupted ecosystems: Rapid temperature changes can disrupt ecosystems and affect species' ability to adapt. Many species have specific temperature ranges in which they can survive and reproduce. When temperatures shift outside these ranges, it can lead to reduced population sizes, altered migration patterns, and mismatches between species and their habitats. Such disruptions can cause imbalances in ecosystems and result in the loss of biodiversity.

SEA-LEVEL RISE AND COASTAL FLOODING

a. Coastal flooding: As global temperatures rise, ice sheets and glaciers melt, and warmer seawater expands. These factors contribute to a rise in sea levels, increasing the risk of coastal flooding. Low-lying coastal areas and densely populated cities are particularly vulnerable. Coastal flooding can damage infrastructure, destroy homes, contaminate freshwater sources with saltwater, and displace millions of people.

b. Saltwater intrusion: Rising sea levels also lead to saltwater intrusion into coastal aquifers and freshwater sources. This intrusion threatens drinking water supplies and agricultural land, as high salt content renders water and soil unsuitable for human consumption and crop growth.

c. Displacement and migration: As coastal areas become uninhabitable due to sea-level rise and flooding, mass migrations of people may occur. The resulting displacement can strain resources in receiving areas, exacerbate social tensions, and lead to conflicts over limited resources and land.

EXTREME WEATHER EVENTS

a. Hurricanes and cyclones: Warmer ocean temperatures provide more energy for the formation and intensification of hurricanes and cyclones. This can result in more frequent and severe storms, causing extensive damage to infrastructure, coastal erosion, and loss of life.

b. Droughts and wildfires: Higher temperatures and changes in precipitation patterns contribute to drought conditions, increasing the risk of wildfires. Drier vegetation and prolonged droughts create favorable conditions for the ignition and rapid spread of fires. Wildfires can destroy forests, homes, and livelihoods, release large amounts of CO_2 into the atmosphere, worsen air quality, and have long-term impacts on ecosystems.

c. Flooding and heavy rainfall: Climate change can disrupt rainfall patterns, leading to more intense and prolonged rainfall events. This increases the risk of flash floods, river flooding, and landslides. Flooding damages infrastructure, destroys crops, contaminates water sources, and poses risks to human safety. It can also facilitate the spread of waterborne diseases.

CHANGES IN PRECIPITATION PATTERNS

a. Water scarcity: Climate change affects rainfall patterns, leading to increased water scarcity in some regions. Changes in precipitation, combined with higher evaporation rates due to rising temperatures, can result in reduced water availability for human consumption, agriculture, and industry. Water scarcity can trigger conflicts, hinder economic development, and lead to societal instability.

b. Agricultural impacts: Changes in precipitation patterns can disrupt agricultural systems. Reduced rainfall and increased droughts can result in lower crop yields, reduced agricultural productivity, and food shortages. Conversely, intense rainfall events can lead to soil erosion, waterlogging, and crop damage. These impacts can have significant consequences for food security, nutrition, and global food prices.

ECOSYSTEM DISRUPTIONS AND BIODIVERSITY LOSS

a. Habitat loss: Climate change affects habitats critical to various species. Rising temperatures, changing rainfall patterns, and other climate-related factors can lead to habitat loss, such as the destruction of coral reefs, mangroves, and forests. These ecosystems provide numerous benefits, including storm protection, water filtration, carbon sequestration, and habitat for countless species. Their loss can disrupt ecological balance and impact human well-being.

b. Species extinction: Climate change is a major driver of species extinction. Many species struggle to adapt quickly enough to changing conditions, especially those with limited mobility and specialized habitat requirements. Rapid changes in temperature, altered precipitation patterns, and disruptions in ecosystems can lead to increased extinction rates. The loss of species disrupts food chains, reduces biodiversity, and weakens ecosystems' resilience to future challenges.

Addressing and mitigating these hazardous impacts requires urgent and comprehensive action to reduce greenhouse gas emissions, adapt to changing conditions, protect vulnerable communities, and preserve ecosystems. By taking proactive steps, we can work toward a more sustainable and resilient future for humanity and the planet.

Social dimensions of climate change

Climate change is deeply intertwined with global patterns of inequality. The poorest and most vulnerable people bear the brunt of climate change impacts, yet contribute the least to the crisis. As the impacts of climate change mount, millions of vulnerable people face disproportionate challenges in terms of extreme events, health effects, food, water, and livelihood security, migration and forced displacement, loss of cultural identity, and other related risks.

VULNERABILITY THEORY AND CLIMATE CHANGE

Vulnerability theory provides a framework for understanding how social factors interact with environmental changes, such as climate change, to create differential impacts on individuals and communities. It emphasizes that vulnerability is not solely determined by exposure to climate hazards but is also shaped by social, economic, and political factors that influence a community's capacity to cope and adapt.

For example, let's consider a coastal community that relies heavily on fishing for its livelihood. Climate change-induced sea-level rise and ocean acidification can negatively impact fish stocks, leading to a decline in catches and income for fishermen. However, vulnerability theory highlights that vulnerability to these impacts is not uniform across the community.

Factors such as poverty, lack of access to education and healthcare, limited infrastructure, and social marginalization can compound the vulnerability of certain individuals and groups within the community. For instance, marginalized groups may lack the financial resources to diversify their livelihoods or relocate to less vulnerable areas. Lack of access to education and information may limit their ability to adapt to changing fishing practices or explore alternative livelihood options. Furthermore, social inequalities may restrict their access to decision-making processes, making it difficult to influence policies that could address their specific needs and concerns.

Applying vulnerability theory in this context allows policymakers and researchers to identify and address the underlying social determinants of vulnerability, rather than solely focusing on the environmental changes themselves. By understanding the complex social dynamics at play, interventions can be designed to enhance the adaptive capacity of vulnerable groups, promote social equity, and reduce disparities in climate change impacts.

Example: The Pacific Island nations

The Pacific Island nations serve as a poignant example of the social dimensions of climate change. These island nations face unique challenges due to their geographical location, small land areas, and heavy reliance on natural resources for livelihoods and cultural identity. Climate change impacts, such as rising sea levels, increased frequency of extreme weather events, and ocean acidification, pose significant threats to their survival and well-being.

These nations are at the forefront of experiencing climate change impacts, including saltwater intrusion into freshwater supplies, damage to coastal infrastructure, and loss of arable land. The social dimensions of climate change are particularly pronounced in these contexts, with indigenous communities facing displacement, loss of traditional livelihoods, and erosion of cultural heritage.

The Pacific Island nations have been proactive in advocating for global climate action and highlighting the need for social justice in climate change responses. They emphasize the importance of recognizing their unique vulnerabilities and the impacts of climate change on their social fabric, cultural identity, and overall well-being.

By incorporating vulnerability theory and examining real-world examples such as the Pacific Island nations, we gain a deeper understanding of the social dimensions of climate change. This approach highlights the interconnectedness of social, economic, and political factors in shaping vulnerability and underscores the need for inclusive and equitable responses to climate change.

Addressing the social dimensions of climate change requires interdisciplinary collaboration, policy interventions that promote social justice, and the empowerment of vulnerable communities. By integrating social considerations into climate change adaptation and mitigation strategies, we can strive toward a more equitable and resilient future for all.

Paying off for the past to ensure safety for future generations

Climate change is a critical global challenge that has gained significant attention due to its profound consequences on ecosystems, societies, and the well-being of future generations. The urgency to address this issue has given rise to the concept of 'paying off for the past to ensure safety for future generations,' which emphasizes the need to acknowledge historical responsibility and make the necessary investments and sacrifices today to safeguard the sustainability and welfare of future populations. This concept can be further understood through scientific reasoning and the recognition of the various risks posed by climate change.

One of the primary reasons for adopting the 'paying off for the past' approach is the recognition that human activities, particularly the burning of fossil fuels, have been the main driver of the increase in greenhouse gas emissions and subsequent global warming. The excessive accumulation of greenhouse gases in the atmosphere traps heat and disrupts the earth's climate system, leading to rising temperatures and myriad adverse impacts. Scientific evidence indicates that the earth's average surface temperature has already increased by approximately 1°C since the pre-industrial era. By acknowledging historical responsibility, we accept that past emissions have contributed to the current state of climate change and recognize the need to take action to mitigate further warming.

Addressing the climate crisis requires a comprehensive approach that encompasses both mitigation and adaptation strategies. Mitigation involves reducing greenhouse gas emissions to limit the extent of climate change. This can be achieved through the implementation of sustainable practices in various sectors such as transportation, agriculture, and industry. By reducing emissions, we can slow down the rate of global warming and minimize the risks associated with climate change.

Moreover, transitioning to renewable energy sources plays a crucial role in ensuring a safer and more sustainable future. Renewable energy, such as solar, wind, and hydropower, offers a cleaner alternative to fossil fuels, significantly reducing greenhouse gas emissions and air pollution. By embracing these sustainable energy options, we can reduce our reliance on finite and polluting resources, mitigating climate change while also improving air quality and public health.

In addition to mitigation efforts, adapting to the changing climate is essential to minimize the negative impacts on ecosystems and societies. Climate change has already resulted in rising sea levels, increased frequency and intensity of extreme weather events, and loss of biodiversity. These changes pose significant risks to coastal communities, agriculture, water resources, and vulnerable ecosystems.

By acknowledging the historical responsibility, we accept the need to invest in adaptive measures that enhance resilience and protect communities and natural systems from the impacts of climate change.

Investing in adaptation measures involves initiatives such as developing climate-resilient infrastructure, implementing effective land-use planning, and enhancing early warning systems for extreme weather events. These measures aim to minimize the vulnerabilities of communities and ecosystems, enabling them to withstand and recover from climate-related shocks and stresses.

By taking action now to mitigate greenhouse gas emissions and adapt to the changing climate, we strive to secure a more stable and livable environment for future generations. This approach recognizes the intergenerational equity and responsibility we have toward future populations. Failure to act would lead to escalating climate-related risks, including more frequent and severe heatwaves, droughts, floods, and storms, along with the associated consequences for human health, food security, and economic stability.

Furthermore, safeguarding the well-being and sustainability of future generations requires a collective effort that extends beyond individual actions. It necessitates international cooperation, policy interventions, and financial investments in sustainable technologies and practices. Governments, businesses, civil society organizations, and individuals all have a role to play in driving the necessary changes to tackle climate change effectively.

In conclusion, the concept of 'paying off for the past to ensure safety for future generations' underscores the urgency of taking action now to address the climate crisis. By acknowledging historical responsibility and making the necessary investments and sacrifices, we can minimize the negative impacts of climate change and shape a safer and more resilient future. Mitigating greenhouse gas emissions, incorporating renewable energy sources, and adapting to the changing climate are all vital components of this approach. It is through these concerted efforts that we can secure a more stable and livable environment, safeguard the well-being of future generations, and ensure a sustainable future.

IV

Climate Ideology

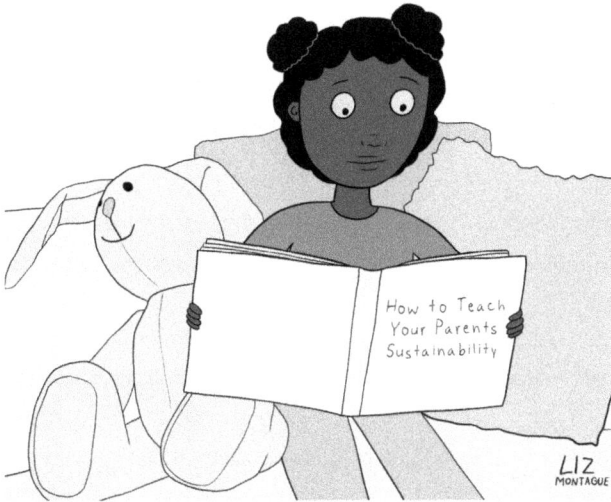

Environmental management

The principles of environmental management have received greater attention in the last few years, particularly as the climate challenge has become a more pressing global concern. The management of climate change is a subset of the broader discipline of environmental management which itself falls under the even broader set of management. Environmental management is inherently opposed to the elements of traditional business management which minimize or ignore the tenets of natural resource management.

In its most idealized form, environmental management adopts the practices of established business management and introduces practical connections between organizations, economic growth models that take account of natural capital while also attempting to strengthen social understanding of this relationship. The alignment of economic growth with natural capital stocks and a greater understanding from the community about this relationship is a foundational cornerstone of environmental management. Environmental management is not as concerned with the solutions as it is with the adoption of appropriate practices.

While this alignment is viewed as the ultimate expression of environmental management principles (indeed, in such form, there would be no real distinction between environmental management and management per se), many have dismissed it as merely an act of lip service which seeks to maintain business-as-usual arrangements. Derogatory terms such as 'hot air' and 'greenwashing' have become common in recent years. More cynical activists might even describe environmental management as essentially an arm of typical business management practice which is not sufficiently concerned with natural stocks and only deals in incremental and defensive concessions to growing public discourse on environmental matters.

Whatever views one holds, the intensity and growing publicity of the climate change debate has inadvertently elevated environmental management principles to significantly higher platforms than ever before. As the climate change 'tail' wags the environmental management 'dog,' it is clear that the traditional, linear business models are set to adopt more circular practices as a nod to climate issues. Ultimately, however, the prudence of considering more holistic issues is likely to prevail over the current focus climate change and businesses will find themselves once more in a 'new new normal.' It is not clear what the 'new new normal' will look like, other than the likelihood it will embed environmental degradation consequences much more clearly into business operating models. Ultimately, environmental management relates to the manner in which businesses and other organizations deal with environmental aspects (i.e., elements or resources of the natural environment).

Environmental management principles

Environmental management is underpinned by decision making, and, more specifically, decision making which is likely to have an impact on the natural environment. Decision making is, in turn, often predicated on principles. Principles are perhaps more important in the field of environmental management than other fields. Their generic form makes them more acceptable in political discourse, and in this sense they strongly contribute to consensus much more readily than various forms of binding instruments such as laws. Additionally, they lend themselves to a wider coverage of implementation across geographies and economic sectors.

There is no single directory of environmental management principles. Unsurprisingly, there are numerous classifications of these principles. It is important to note that principles can be interpreted differently across different jurisdictions. The aim of this section is not to provide the reader with a detailed analysis of various principles; rather, the section introduces the reader to an array of the major principles in order to provide a framework for understanding key guiding theories that have influenced and continue to influence modern-day policy and corporate arenas.

NO-HARM RULE

The no-harm rule is a fundamental principle of customary international law which compels a state to prevent, reduce, and control the risk of environmental harm to other states. This rule gained prominence after a 1941 tribunal ruling of a cross-border case involving a Canadian zinc smelter whose industrial emissions resulted in adverse air pollution effects across the border, in the United States. The rule was subsequently incorporated into numerous policy efforts and it was expanded in the 1972 United Nations Stockholm Declaration on the Human Environment into an overarching statement of responsibility to contain environmental damage within national borders. Principle 21 of the declaration asserts that 'States have, in accordance with the Charter of the United Nations and the principles of international law, the sovereign right to exploit their own resources pursuant to their own environmental policies, and the responsibility to ensure that activities within their jurisdiction or control do not cause damage to the environment of other States or of areas beyond the limits of national jurisdiction.' The second part of Principle 21 is important in that it mentions state responsibility to prevent environmental harm to other states through activities carried out on ships or aircraft registered within a state. It therefore also includes protection of the global commons, i.e., the notion of protection of areas beyond the jurisdiction of any state.

Principle 21 is also repeated in the 1992 Rio Declaration on Environment and Development (Principle 2), the Convention on Biological Diversity – CBD (Article 3), and the United Nations Framework Convention on Climate Change – UNFCCC (in recital 8 of its preamble).

SUSTAINABLE DEVELOPMENT

In 1987, the World Commission on Environment and Development (WCED), published the Report of the World Commission on Environment and Development: Our Common Future. The Brundtland Report (after the Commission's chairwoman, Gro Harlem Brundtland) is widely credited as the source of the most acceptable definition of sustainable development. It states that global environmental problems are an outcome of divergent ends of economic development – significant poverty of the geopolitical south and unsustainable economic growth patterns of the geopolitical north. It goes on to define sustainable development as 'development that meets the needs of the present without compromising the ability of future generations to meet their own needs.'

The principle of sustainable development provides the foundation of national and international efforts of environmental protection, in the absence of stronger imperatives for harm prevention. The absence of enforceable burden on parties to demonstrate that, in fact, development has been 'sustainable' has led to some calls to consider sustainable development as a more abstract concept than a principle, something closer to a value. Nonetheless, the principle of sustainable development is often considered as a family of sub-principles which should be viewed together as principles of sustainable development.

The precautionary principle: The precautionary principle is one of the most well-known principles of sustainable development. It is based on the idea of taking anticipatory action in order to prevent environmental harm, even if the unequivocal scientific evidence is not yet available. It is a more risk-conscious approach to environmental management, and has its origins in Germany. In 1972, West Germany amended its constitution to manage the causes of acid rain, which had led to environmental harm, such as damage to German forests. The constitutional amendment progressed into a federal statute which had two stated goals: 'to protect people, animals, plants, and other things from harmful environmental effects' and 'to take precautions against the occurrence of harmful environmental effects.' While the West German air pollution standards were not initially more stringent than many industrialized countries, they were progressively strengthening under the banner of *Vorsorgeprinzip*, the precautionary principle. West German policymakers began applying stringent restrictions on entire classes of chemicals – and not just specific chemicals, as other countries did – even where scientific proof of harm-causing capabilities was not yet available.

In 1992, the precautionary principle gained its moment of global recognition when over 150 countries adopted the Rio Declaration at the Earth Summit. This was the biggest environmental agreement at the time, and formed the basis for many subsequent pollution and climate treaties. Principle 15 of the summit's declaration states that 'In order to protect the environment, the precautionary approach shall be widely applied by States according to their capabilities. Where there are threats of serious or irreversible damage, lack of full scientific certainty shall not be used as a reason for postponing cost-effective measures to prevent environmental degradation.'

The principle was also codified into the high profile environmental multilateral agreements at the summit: namely, the Convention on Biological Diversity (CBD) and the United Nations Framework Convention on Climate Change (UNFCCC).

In practical applications, once the principle is invoked it does not mean that 'risky' activities do not proceed further along the decision-making trail. Rather, decisionmakers are to be guided by the spirit of careful consideration to avoid significant, irreversible environmental harm where practically possible, and assess the pros and cons of other options through a risk lens.

Inter-generational equity, intra-generational equity & differentiation: In a most simplistic viewpoint, it could be argued that the ideals of environmental protection are encapsulated within the principle of inter-generational equity, while the concept of development is encapsulated within the principle of intra-generational equity, and that the notion of fairness is at the core of the differentiation principle.

The principle of inter-generational equity holds that every human generation has the right to enjoy resource benefits from past generations and, equally, the responsibility to retain these resource benefits for future generations. A concept of fairness among generations is at the core of this principle, which is, of course, the most famous element of the previously discussed Brundtland Report which conceptualized it as 'development that meets the needs of the present without compromising the ability of future generations to meet their own needs.' Equity considerations were documented even earlier in the form of the 1972 Stockholm Declaration, whereby Principle 2 asserted the need for the world's resources to be protected for the benefit of present and future generations. More recently, the Paris Agreement calls for multilateral efforts to be made on the basis of equity and sustainable development considerations.

At the core of the principle is the idea that the overall health of the environment be at least maintained, if not improved, for future generations. This is more than simply a long-term assessment of impacts. It entails a more significant and reflective assessment of a locale, the cumulative effects of environmental harm, and the expectation of a 'net zero' loss of environmental amenity value.

Related to inter-generational equity is the idea of intra-generational equity, which relates to ideas of fairness and justice across the communities within the current generation. In real-world application, the notion of intra-generational equity has generally been embodied most prominently within poverty eradication initiatives, such as ensuring fair access across communities to social services such as education and healthcare. Other poverty eradication initiatives within the intra-generational equity scope are the reduction of income inequality and the promotion of gender equality and empowerment of marginalized groups.

Common but differentiated responsibilities and respective capabilities (CBDR-RC) is a principle which speaks to fairness in contributions to address past harm and capability to respond to climate change. The principle is enshrined in the 1992 UNFCCC treaty, which states that 'the global nature of climate change calls for the widest possible cooperation by all countries and their participation in an effective and appropriate international response, in accordance with their common but differentiated responsibilities and respective capabilities and their social and economic conditions.' The Rio Declaration also notes that 'the least developed and those most environmentally vulnerable, shall be given special priority.' While CBDR-RC has been a principle that has guided decision-making processes, it has also been a source of consternation in the history of UN multilateral climate negotiations.

These three principles are sympathetic to equity issues across past, current, and future generations. Inter-generational equity has a clear focus on preservation of future rights; intra-generational equity focuses on the distribution of resources and opportunities within a generation. CBDR-RC is chiefly concerned with past injustices and methods of reflecting variations in capabilities of states to attend to such injustices.

Polluter pays & user pays principles: The polluter pays principle (PPP) was indirectly incorporated in Principle 22 of the 1972 Stockholm Declaration and was significant because it called for the inclusion of provisions in international environmental law to address liability and compensation for the victims of pollution. This principle holds that the party which causes pollution should bear the costs of the damage caused and associated remedial actions. It has been such a pillar of environmental management that it has been contained in numerous multilateral environmental agreements since Stockholm. In 1992 the principle was adopted in Principle 16 of the Rio Declaration. It attempts to mitigate environmental harms by placing a cost on pollution. How does it attempt this obligation? It aims to compel parties that benefit from the unaccounted externalities of pollution to internalize those costs by paying for the pollution costs of reduction, containment, disposal, or any other mitigation measures. The principle has mainly been implemented through command-and-control approaches (such as prohibition of certain activities without appropriate licenses which usually impose conditions that may require pollution control measures, for example) but can also be applied through market-based mechanisms, e.g., for the development and introduction of environmentally sound technologies and products.

In practice, the principle is implemented generally through some sort of license fees paid by the party that undertakes the polluting activities. There are some issues with this practice and they relate mainly to determining viable thresholds on which to establish the value of compensation. License fees rarely bear out the true cost of the pollution itself. Instead, they tend to primarily reflect administration costs. An exception to this practice is a form of load-based licensing, which is a feature of some Australian jurisdictions, for example. The load-based licensing (LBL) scheme of the state government of NSW 'sets limits on the pollutant loads emitted by holders of environment protection licenses and links license fees to pollutant emissions.' In so doing, it aims, among other objectives, to provide incentives to reduce the load of pollutants emitted based on the polluter pays principle (NSW EPA Link). Essentially load-based licensing sets fees not so much on volume of waste generated but on the risks of environmental harm.

The user pays principle (UPP) is a variation of the PPP that requires the user of a natural resource to bear the cost of running down natural capital. The UPP is a complement to the PPP, and it attempts to internalize production and consumption externalities by ensuring that the full lifecycle of inputs is priced to reflect the function and usage of the goods. PPP tends to focus on responsible accountability and management of waste, whereas UPP tends to focus on improving pricing structures, often leading to the end users paying fees for use.

While the idea of charging users a fee to enjoy the benefits of environmental resource value is relatively innocuous, the actual methods to assign a pricing value to environmental products and services is particularly difficult. Various tools have been employed to affect such pricing assignments. At the easier end of the spectrum are direct-use assessments which look at valuation of physical consumption of resources, such as water bodies. At the more complex end of the spectrum are the non-use valuations, such as non-use values of landscapes, which are harder to quantify and, in some cases, even attract outright rejection from some stakeholders.

Biodiversity offsets are a form of UPP that are gaining popularity. Such offsets require land developers to pay an opportunity cost for the development of land in the form of procurement of alternative tracts of land or funding of research grants with the overall aim of offsetting the loss of biodiversity value or landscape amenity value for the community.

PPP and UPP play a significant role in environmental management, acting as deterrents and directing accountability for environmental harm.

The principles of environmental management provide a robust foundation for sustainable decision making and action. By integrating these principles into corporate and policy-making approaches, we can foster a harmonious relationship between human activities and the natural environment. These principles are ultimately aimed to guide us toward a more sustainable and resilient future.

Managing environmental challenges

The principles and implementation of environmental management have advanced over the years through the influences of multidisciplinary, cross-cultural, and technological developments. Environmental decision making has increasingly favored – indeed, required – the use of multidisciplinary platforms to resolve issues. It is not unusual to see the involvement of seemingly unrelated professions and parties in the decision-making process. As decision makers make more efforts to integrate various disciplines, a key issue remains unresolved, and that is how to deal with scientific uncertainty (the heart of the precautionary principle).

For generations, environmental management issues have been, and continue to be, characterized by the multiplicity of competing interests and stakeholders. This interplay of interests was a feature of Rachel Carson's *Silent Spring*, published in 1962 and is still a feature of climate negotiations in the few years remaining to the maturity of the Sustainable Development Goals in 2030.

Once an environmental issue is framed and publicized, different stakeholders embrace their own views on the issue. For the ultimate decision makers, the multiplicity and lobbying of these stakeholders presents a confounding reality: It is nearly impossible to reach a solution which satisfies the 'right' number of stakeholders, protects the environment, addresses societal justice issues, and promotes an 'acceptable' degree of continued economic development.

The environmental management principles discussed in the previous section should provide the appropriate framework for decision makers and other stakeholders to frame the issues and deliberate on their eventual resolution. Additionally, a wide range of climate ideologies is constantly exerting pressures on decision-making infrastructure. These ideologies include the following, to name a few:
- Climate skepticism/denial
- Climate realism/pragmatism
- Environmentalism
- Climate justice
- Climate activism
- Techno-optimism

It is becoming clearer through the development of climate negotiations in the international arena, and also in the maturing implementation of domestic environmental schemes, that more acceptable decisions can best be discharged through more robust governance structures. The benefits of such structures have been discussed by Godden et al. (2019) and include the following:
- Reconnection with a more fundamental concept of participation which engages those potentially affected to have a say in decision-making processes

- Securing greater participation and loyalty to environmental decisions as a result of their greater engagement of stakeholders in the first place
- Creating new channels for accessing information, managing compliance obligations, and enforcing rules as per broad-based commitment
- Enhancing the adaptivity and flexibility of existing and emerging regulatory systems
- Meeting international standards for increased participation in sustainable development ideals and environmental laws

A governance process would still have significant challenges to deal with. For instance, given the broad base of stakeholders, how should the contemporary discussions on climate change be best addressed? Again, Godden et al. (2019) hint at the difficulty by posing this question: Does the urgency of a perceived environmental problem necessitate a restriction of full participation in order to expedite the process of decision making, and how should conflicting views on such problems be reconciled?

Perhaps the single greatest challenge of environmental management issues does not lie in determining the scientific basis for decisions; neither does it lie in making decisions in the face of scientific uncertainty; rather, it is how to practically entertain the views of multiple and competing interests and synthesize them into practical solutions. Ultimately, a governance system is required which can manage across a number of troublesome issues, such as (once more leaning on the work of Godden et al.) conflicting policy objectives, socio-economic justice considerations, variation of environmental phenomena, and high levels of scientific uncertainty.

Climate negotiations

Concern for the environment, spurred in part by events such as the seminal publication of *Silent Spring* by Rachel Carson in 1962, provided the basis for the eventual globalization of environmental policy from the 1970s. The 1972 United Nations Conference on the Human Environment in Stockholm was the first world conference to make the environment a major issue. The participants adopted 26 principles for management of the environment, including the Stockholm Declaration and Action Plan for the Human Environment. Importantly, the conference also demanded the creation of international institutions and mechanisms. This would form the basis for the formation of the institutional framework and multilateral environmental agreements that have become a defining feature of environmental policy in the subsequent decades, headlined by climate negotiations.

Climate negotiations refer to international discussions and agreements aimed at addressing climate change and reducing greenhouse gas emissions. These negotiations involve governments, intergovernmental organizations, and various stakeholders, such as non-governmental organizations (NGOs), businesses, and scientists. The primary objective of climate negotiations is to develop and implement effective policies and measures to mitigate climate change and adapt to its impacts.

The list below is merely a scan of key climate negotiations and outcomes in the decades since the resolutions of the Stockholm Conference in 1972:

1. United Nations Framework Convention on Climate Change (UNFCCC): Adopted in 1992, the UNFCCC is an international treaty to address climate change; it serves as the foundation for global climate negotiations since it includes virtually all nations. It is tasked with the objective of working toward stabilizing greenhouse gas concentrations to prevent dangerous human interference with the climate system.

2. Conference of the Parties (COP): The UN climate change conferences are the official meetings of the Conference of the Parties (COP). The UNFCCC is the global voice for climate change and one of the three major UN treaties known as the Rio Conventions, alongside the UN Convention on Biological Diversity and the UN Convention to Combat Desertification. The COP is the highest decision-making body of the UNFCCC. It convenes annually, bringing together representatives from almost all countries. COP meetings serve as a platform for negotiating and reviewing climate agreements, including the Paris Agreement.

3. Kyoto Protocol (COP3): The Kyoto Protocol, adopted in 1997, was an international treaty linked to the UNFCCC. It was the first global climate agreement to call for country-level reductions in greenhouse gas emissions in industrialized nations, as it set binding targets for industrialized countries. The agreement acknowledged that developing countries should not bear the same responsibility for limiting emissions as industrialized economies. The agreement established several flexibility mechanisms that allowed developed countries to meet part of their climate commitments through investments in emission reductions in developing nations. It has largely been replaced by the Paris Agreement.

4. Paris Agreement (COP21): The Paris Agreement, adopted in 2015 and entering into force in 2016, is another landmark international agreement under the UNFCCC. It is a legally binding international agreement that has set a goal of limiting global warming to below 2°C above pre-industrial levels and pursuing efforts to limit the temperature increase to 1.5°C. Building on the efforts of the UNFCCC and the Kyoto Protocol, this agreement requires countries (both developed and developing) to submit nationally determined contributions (NDCs) outlining their efforts to reduce emissions and adapt to climate change.

5. Nationally determined contributions (NDCs): NDCs are bottom-up commitments made by countries toward the mitigation and adaptation goals of the Paris Agreement. Each country sets its own targets, policies, and actions to address climate change based on its national circumstances. The NDCs are reviewed and updated periodically to enhance ambition and strengthen global climate action. Countries are legally required to submit new NDCs every five years to demonstrate increased ambition and progress toward the Paris goals.

6. COP26: The 26th UN Climate Change Conference of the Parties (COP26) was held in Glasgow, Scotland, in 2021. One of its main outcomes was the Glasgow Climate Pact, which includes agreements to strengthen efforts toward climate change

resilience, ramp climate finance (including a reaffirmation of the COP15 $100 billion-dollar annual pledge from developed to developing countries) enhance adaptation efforts, and finalize the rules for implementing the agreement.

Despite numerous successes in delivering outcomes that reflect integration of disciplines and stakeholder interests, climate negotiations still have significant barriers to overcome in order to deliver truly broad-based solutions to complex issues. They have to cater to global issues while addressing local concerns, often within the context of highly fragmented policy and regulatory environments. They touch topics that are both technical and non-technical as a result, including emission reduction targets, climate finance, technology transfer, adaptation measures, capacity building, and transparency in reporting and verification.

Climate negotiation processes and challenges

Climate negotiations form the basis for consensus in multilateral environmental agreements (such as protocols, conventions, treaties, and others). These negotiations often apply some of the principles of environmental management (discussed earlier) and generally engage stakeholders to bind themselves to obligations to protect specific environmental aspects. Stakeholders typically participate to ensure alignment between these agreements and their own parent organization or constituency policies. Some of the most well-known agreements are:

- United Nations Framework Convention on Climate Change (UNFCCC)
- Paris Climate Agreement
- United Nations Environment Program (UNEP)
- Convention on Biodiversity and the Cartagena Protocol on Biosafety
- Basel Convention on the Control of Transboundary Movements of Hazardous Wastes and Their Disposal
- Convention on International Trade in Endangered Species of Wild Fauna and Flora
- Minamata Negotiations on Mercury

- Montreal Protocol on Substances that Deplete the Ozone Layer
- Stockholm Convention on Persistent Organic Pollutants

The nature of these 'soft law' agreements has changed over the years, as the parties have sought to broaden not just the scope of the agreements themselves but also the participatory process. For example, at the start of the millennium, the international development agenda was encapsulated within the Millennium Development Goals (MDGs) which were focused on reducing various aspects of poverty in developing countries. In 2000, the MDGs established measurable, universally agreed objectives aimed at combatting poverty. The Sustainable Development Goals (SDGs) were born at the UN Conference on Sustainable Development, in Rio de Janeiro, in 2012, and came into force to replace the MDGs in 2015. While the MDGs were prepared by a group of UN experts, the SDGs were the outcome of a very broad consultation process of working groups, civil society organizations, the general public, and other stakeholders. Additionally, while the MDGs focus on eight goals, 21 targets, and 63 indicators, the SDGs comprise 17 goals, 169 targets, and 232 indicators. Also, while the overall scope of the MDGs could be described as social, the SDGs cover economic, social, and environmental areas. Finally, the coverage of the MDGs was developing economies, while the SDGs cover both developing and developed economies.

Climate negotiation processes are structured discussions and procedures followed during international climate negotiations. These processes are designed to facilitate dialogue, decision-making, and the development of agreements among participating parties. Here are some key aspects and challenges of climate negotiation processes:

1. Negotiation tracks: Climate negotiations typically involve multiple tracks or working groups focusing on different aspects of the climate agenda. These tracks often include mitigation (reducing greenhouse gas emissions), adaptation (addressing the impacts of climate change), finance (mobilizing financial resources for climate action), technology transfer, capacity building, and transparency.

2. Consensus-based decision making: Climate negotiations strive to reach consensus among participating parties. This means

that decisions and agreements are made collectively, taking into account the concerns and perspectives of as many parties as possible. As discussed earlier, different parties have their own interests, priorities, and political dynamics, and achieving consensus is difficult.

3. Differentiation: A significant challenge in climate negotiations is the acknowledgment of differentiation between developed and developing countries. Historically, developed countries have been responsible for most greenhouse gas emissions, while developing countries face challenges related to poverty eradication and sustainable development. Negotiations have often revolved around the allocation of responsibilities, financial assistance, and technology transfer between these two groups.

4. Ambition and equity: Climate negotiations require balancing the ambition of emission-reduction targets with considerations of equity. Developing countries have argued that developed nations should take the lead in reducing emissions and providing financial support, given their historical responsibility and greater capacity. However, developed countries often emphasize the need for global participation, and in more recent years, also the need for shared responsibilities.

5. Climate finance: Mobilizing financial resources to support developing countries in their climate actions is a critical aspect of climate negotiations. The provision of climate finance, including adaptation funding and technology transfer, remains a challenge. Negotiations focus on issues such as the scale of funding, transparency, accountability, and ensuring access to finance for the most vulnerable countries.

6. Adaptation and loss & damage: Negotiations on adaptation to climate change and addressing loss and damage (irreversible impacts) are crucial to the success of negotiation events. Determining funding mechanisms, supporting vulnerable countries in building resilience, and addressing issues of liability

and compensation for loss and damage are complex and contentious topics.

7. Transparency and accountability: Ensuring transparency and accountability in reporting emissions, progress on targets, and the use of financial resources are vital for building trust among countries. Negotiations focus on establishing robust monitoring, reporting, and verification systems that are acceptable to all parties.

8. Time-frames and urgency: Climate negotiations face the challenge of aligning long-term climate goals with short-term actions. Urgent and ambitious action is needed to address the climate crisis, but negotiations can be slow and protracted processes. Balancing the need for immediate action with the complexities of decision making and national processes poses a significant challenge.

Climate negotiations face significant challenges in developing effective global agreements. One major hurdle is the differentiation between developed and developing countries. Developing nations demand greater commitments and financial support from developed countries, citing historical responsibility and capacity limitations. Striking a fair balance between countries' responsibilities remains complex.

Balancing ambition and equity poses another challenge. Achieving ambitious emission reduction targets while considering varying national circumstances is difficult. Finding a middle ground that ensures action while acknowledging different levels of development is crucial.

Adaptation and addressing loss and damage are complex topics. Establishing funding mechanisms, supporting vulnerable nations, and addressing liability and compensation are contentious. Striking a balance between support and responsibility is crucial.

Transparency and accountability in reporting emissions, progress, and finance usage are vital but challenging. Developing robust monitoring and reporting systems acceptable to all parties remains an on-going task.

Aligning long-term goals with short-term actions is complex. Urgent action is necessary, but negotiations can be time-consuming due to diverse national processes. Effective diplomacy and sustained engagement are required.

Perhaps one of the most contentious issues is climate finance, which, if it fails to materialize, or creates significant negotiation inertia along the way, could undermine trust between nations. At COP15 in 2009 (Copenhagen), developed economies committed to mobilize US$100 billion in climate finance per year by 2020. At COP21 in 2015 (Paris), developed economies extended that commitment to 2025. At COP26 in 2021 (Glasgow), leaders expressed 'deep regret' at the failure to meet the US$100 billion annual commitment. It is widely agreed that developed economies have failed to reach the annual target despite the renewed commitments of COP29. Additionally, there are growing concerns about the quality of proposed financing arrangements, with contention around issues such as the use of loans rather than grants, differing definitions of 'climate finance,' and, of course, insufficient funding for adaptation initiatives themselves. Yet another trust issue is likely to form over the preference of developed economies to expand the list of climate financers to account for the increasing wealth of some countries traditionally considered as developing economies. This would require a significant shift in the trust of developing economies toward developed economies which, as of this publication, is not near to materializing.

Overcoming these challenges requires continuous effort, negotiation nous, and political will. Regular assessment and improvement of negotiation processes are essential to strengthen global climate governance. By recognizing complexities and collaborating, countries can work toward effective climate agreements.

Negotiation to execution

Bearing in mind that the main goal of the UNFCCC is to achieve stabilization of GHG concentrations in the atmosphere at a level that would prevent dangerous anthropogenic interference with the climate system, it is important to understand the institutional arrangements that have been established to reach this overarching goal. The Convention and the Kyoto Protocol provided provisions for institutional arrangements, such as establishing the Conference of the Parties (COP), Bureau, Subsidiary Bodies (SBs), Secretariat, as well as Ad-hoc Working Groups (AWGs) and limited-membership bodies. These supreme bodies generally convene annually and examine the parties' commitments while also coordinating measures to address climate change effects, mobilize climate finance, and discharge various administration functions.

The member countries (known as 'Parties,' hence the term 'Conference of the Parties') are generally responsible for achieving outcomes of the UNFCCC ('the convention' for purposes of this discussion) and relevant agreements such as the Paris Agreement (or, previously, the Kyoto Protocol). As a result, they engage in negotiations to find ways and means to implement decisions. While the convention and its relevant agreements do not have an official process for establishing negotiation groups, it is conventional practice for parties with common interests to form groups and generally inform the COP Bureau or relevant secretariat accordingly. In many cases, these common interest groupings are based on developing and developed economy interests.

Here is an overview of the typical steps involved in the climate negotiation process, from negotiation to execution:

1. Pre-negotiation: Before negotiations begin, countries often engage in pre-negotiation activities, such as research, analysis, and consultations with stakeholders to develop their positions and priorities. They may also participate in regional or international conferences to exchange ideas and build consensus.

2. Negotiation: The negotiation phase involves formal discussions and meetings where representatives from various countries, known as delegates, come together to negotiate an agreement. These negotiations typically take place within the framework of the convention. The negotiations can span several rounds and may occur at different levels, including annual conferences (e.g., Conference of the Parties, or COP) and subsidiary bodies.

   Key negotiation topics may include:
   • Emission reduction targets: Determining the overall emission reduction goals for each country, or group of countries.

- Adaptation: Addressing measures to help communities and ecosystems adapt to the impacts of climate change.
- Finance: Mobilizing and allocating financial resources to support climate mitigation and adaptation efforts, particularly for developing countries.
- Technology transfer: Facilitating the sharing of environmentally friendly technologies to support climate action.
- Transparency and accountability: Establishing mechanisms to monitor and report progress on emissions reductions and other commitments.

Negotiations can be complex and contentious, as countries often have diverse interests, economic priorities, and varying levels of responsibility for climate change.

3. Agreement and signing: Once negotiations reach a consensus, an international agreement is drafted and presented to the participating countries for signing. The most notable agreement to date is the Paris Agreement, adopted in 2015, which aims to limit global warming to well below 2°C above pre-industrial levels.

4. Ratification and domestic implementation: After signing the agreement, countries need to ratify it through their domestic legislative processes. Ratification demonstrates a country's commitment to abide by the agreement's provisions, and sets the stage for its implementation. Governments may need to enact new laws, policies, and regulations, allocate resources, and establish institutions to implement the agreement effectively.

5. Execution and compliance: Once the agreement is ratified, countries begin executing their commitments. This involves implementing the agreed-upon measures, policies, and actions at the national level to achieve the stated climate goals. Countries are expected to regularly report on their progress and submit national reports to demonstrate compliance with their commitments.

6. Review and revision: Periodic reviews are conducted to assess the collective progress toward meeting the agreement's

objectives. These reviews often take place during international conferences, such as COP meetings, where countries evaluate their efforts, share experiences, and negotiate updates or revisions to the agreement based on emerging scientific findings or changing circumstances.

The climate negotiation process is ongoing and iterative, with continual refinement and adaptation based on new scientific knowledge, evolving priorities, and international cooperation. It requires sustained political will, collaboration, and engagement from governments, civil society, and other stakeholders to address the urgent global challenge of climate change.

V

Sustainability and ESG

"They're an invasive species that will destroy the environment if left unchecked."

ESG investing has the power to slow climate change, increase
the equity of those that have none, and make a profit.

Climate represents a third of ESG. Climate change is the greatest threat to human health, and this includes malnutrition and starvation, increases in infectious disease, extreme weather events such as floods and droughts, increased frequency of wildfires, loss of livelihoods (and ensuing social unrest), forced migration from affected areas (which can lead to immigration issues as well as increased crime rates), and loss of human life.

The risks associated with climate change have a massive effect on businesses. For example, severe weather events, such as floods and droughts, can damage physical assets such as property and equipment; changes in temperature, rainfall, and natural disasters like hurricanes and typhoons can impact agricultural yields (this can lead to increased food prices and decreased wages for agricultural workers); increases in the frequency of wildfires can reduce timber supplies, which impacts profit margins.

It means that any actions taken to reduce climate change should also be viewed as opportunities for future growth. The fight against climate change requires a large amount of time, money, and other resources. Businesses can look at this as an opportunity to get ahead by investing in sustainable energy sources, by increasing recycling efforts on their part, or even reducing the environmental impact of their material suppliers. ESG considerations should be included in a business's long-term plans and goals, because it offers an opportunity for growth.

What is sustainability?

Sustainability is the ability to meet our requirements without jeopardizing future generations. Resources needed also include social and economic ones.

'There is no such thing as "away".
When we throw anything away, it must go somewhere.'
– Annie Leonard, proponent of sustainability

It is not only a concern for the environment; economic growth and social equity are also tied to it. Preservation of a shared environment for future generations is a major problem.

Sustainability plans deal with environmental factors, like:

- Conservation of the environment
- Maintenance of renewable fuel sources
- Decrease in CO_2 emissions

In short, it fosters innovation and preserves our way of life. It also protects our natural environment and human and ecological health.

Sustainability business is an approach to doing business that considers the needs of both current and future generations. It balances economic growth, environmental protection, and social progress. It does not compromise the ability of future generations to meet their own needs.

This implies that companies must consider how their activities may affect global resources. Be it the environment or civil society. This entails cutting back on waste and emissions.

It raises energy efficiency and encourages sustainable supply-chain practices. It also involves making investments in renewable energy sources.

Businesses must ensure lowering any harmful environmental consequences while generating value for everyone. Be it stakeholders, employees, consumers, or shareholders.

Sustainable business practices aim to end poverty, combat inequality, and halt climate change. To do this, companies must ensure that their supply chain supports environmental objectives. Such as tackling global warming, preserving underwater life, and preserving terrestrial life.

'The activist is not the man who says the river is dirty.
The activist is the man who cleans up the river.' – Ross Perot

They could check their surroundings to determine how they can contribute to good health and well-being, high standards in education, access to clean water and sanitation, and the development of sustainable towns and cities.

These objectives all complement one another to yield a beneficial outcome.

Emerging sustainability issues

As the world finds itself at the crossroads of particularly consequential times in the 21st century, global policymakers are grappling with the implications of significant environmental, social, and economic policy decisions. As they exchange ideas on solutions, the needs of future generations have become central to these decision-makers. The three areas of environmental, social, and economic decisions are inextricably linked to each other. Having said that, of the environmental issues in discussion, the most pressing to hit policy agendas are those related to climate change, energy transitions, food security, sustainable urban development, social inequality, biodiversity conservation, water resource management, and the shift toward a circular economy. These issues require coordinated international cooperation, innovative solutions and capacity building initiatives.

CLIMATE CHANGE

Climate change has emerged as the focal point of humankind's most pressing issue of modern times. Rising global surface temperatures and changes in precipitation patterns are thought to be behind the increasing frequency and intensity of extreme weather events. Changes of a few degrees in average surface temperatures of the planet can lead to perilous changes in climate and weather.

Climate impacts are often classified as primary impacts and secondary impacts. Primary impacts are those that are directly caused by a climate hazard and are generally biophysical in nature (e.g., sea level rise, intensified tropical storms and bushfires). Secondary impacts are those that result from primary impacts and tend to be socioeconomic (e.g., damage to infrastructure).

Changes in temperature and precipitation will almost certainly be the cause of the expected intensification of phenomena such as hurricanes, wildfires, heatwaves, and floods. Climate change is already leading to the intensification of extreme weather events. Even though many impacts have been documented for many years, some issues are emerging as signposts for a series of growing problems, chief among them are the reality of water scarcity and the urban issue of the heat island effect.

Climate change is already threatening access to clean water. Lack of potable water and adequate sanitation is already a major problem around the world and a major human health problem. Freshwater sources are expected to be further compromised by climate change and this will be exacerbated by growing urban populations. The result will be an urban population that struggles to meet its water requirements with an already struggling water infrastructure network which will only worsen in the coming decades. The severity of the water scarcity issue will be location-specific due to differences in freshwater sources, population characteristics, and institutional and financial capacity.

Climate change impacts on urban heat are another serious emerging issue as the urban global population and associated infrastructure continues to grow. Urban populations generally experience higher average temperatures and more intense heat extremes than non-urban populations as a result of the materials used in buildings and infrastructure, which are known to absorb more heat compared to natural materials. An urban heat island is a municipal area that is significantly warmer than the non-urban areas that enclose it. As urban areas trap more heat than non-urban areas, they contribute to human health problems and also impact economic activity, the natural environment, and urban infrastructure. The issue of human health impacts has gained more attention because urban heat islands typically hold more pollutants in the air (vehicular and industrial emissions), which are unable to escape into wider spaces. Urban heat and the urban heat island effect are having measurable effects on heat-related impacts of climate change in urban areas, and these effects are already taking a toll on vulnerable segments of the population, such as children and senior citizens.

FOOD SECURITY

The impacts of climate change on food insecurity are complex but the relationship is very clear. It is also clear that climate change will make it harder for governments to ensure the adaptability of their food supply to these climactic changes. The intensification and increasing frequency of extreme weather events will damage crops, livestock, and fisheries but it will also damage infrastructure inputs to the food production system, including infrastructure related to the production, transport, processing, packaging, storage, retail, consumption, and loss and waste of food.

The IPCC has already commented on the impact of climate change on food security. As part of its Sixth Assessment Report a working group from the IPCC published the Climate Change 2022: Impacts, Adaptation and Vulnerability report which assesses the impacts of climate change, looking at ecosystems, biodiversity, and human communities at global and regional levels. In the report, which highlights that some natural and human systems have already been pushed beyond their capacity to adapt to climate change, the IPCC makes the following observation about food security if global average temperatures rose beyond 2°C:

'…At 2°C or higher global warming level in the mid-term, food security risks due to climate change will be more severe, leading to malnutrition and micro-nutrient deficiencies, concentrated in Sub-Saharan Africa, South Asia, Central and South America and Small Islands. Global warming will progressively weaken soil health and ecosystem services such as pollination, increase pressure from pests and diseases, and reduce marine animal biomass, undermining food productivity in many regions on land and in the ocean.'

It is worth noting that global food insecurity has increased in recent years. In fact, the 2024 edition of the World Migration Report points out that:

'…global food insecurity has dramatically increased during the last 10 years, partially as a result of changes in the climate, but also due to an increase in conflict (both frequency and intensity) and economic slowdowns, compounded by the effects of the COVID-19 pandemic. Direct impacts of climate-related events on food security are most visible with sudden-onset disasters (such as hurricanes or floods), which tend to destroy community infrastructure or damage agricultural landscapes.'

Although less visible and 'dynamic,' slow-onset climate events such as drought or rising sea levels) contribute significantly to food insecurity by compelling populations to make particularly tough decisions about their wellbeing, such as the need to relocate.

Food insecurity is a complex issue and it is not easy to separate climate factors from other contributors such as social, economic, and political factors. Food systems interact with numerous other non-climate factors and therefore when assessing the vulnerability of communities to climate-induced food insecurity, it is important to undertake a more holistic approach.

Increased frequency and severity of extreme weather events, coupled with increasing temperatures and changing precipitation patterns are already playing with the yields of some crops like maize (corn). In lower-latitude regions yields have been affected negatively by observed climate changes, while in higher-latitude regions yields have been positively affected.

The 2019 'Climate Change and Land: An IPCC Special Report on climate change, desertification, land degradation, sustainable land management, food security, and greenhouse gas fluxes in terrestrial ecosystems' notes that food security will be increasingly affected by projected future climate change. It highlights that 'while increased CO_2 is projected to be beneficial for crop productivity at lower temperature increases, it is projected to lower nutritional quality (e.g., wheat grown at 546–586 ppm CO_2 has 5.9–12.7% less protein, 3.7–6.5% less zinc, and 5.2–7.5% less iron). Distributions of pests and diseases will change, affecting production negatively in many regions. Given increasing extreme events and interconnectedness, risks of food system disruptions are growing.' Climate-related food insecurity is a major human issue of our times and may prove to be an even more significant issue in generations to come.

CLIMATE MIGRATION

Although there is nothing new about migration induced by environmental factors, climate change has triggered growing internal and international migration and displacement. The impacts can be clear, such as where drought displaces entire communities. Quite often, however, the impacts are harder to directly trace to climate change.

Although common perceptions of environment-induced migration tell stories of mass movements, the evidence suggests that most climate migration is within borders, temporary in nature, and very specific to the adaptive capacity of affected communities.

Latin America, South Asia, and sub-Saharan Africa have been commonly identified as the most vulnerable to the effects of climate change and could, therefore, likely experience significant movement of people from both internal and cross-border migration.

In its 2018 Groundswell: Preparing for Internal Climate Migration report, the World Bank estimated in its pessimistic scenario that without urgent national and global climate action, Sub-Saharan Africa, South Asia and Latin America could see more than 140 million people move within their respective borders by 2050. Its slightly less pessimistic scenario has this figure at just over 70 million internal climate migrants. In all scenarios of this assessment, the poorest people and the poorest countries are worst hit by climate migration.

Climate change impacts such as sea level rise also increase the probability of migration under duress. As a result of existing inequalities, vulnerable members of the community have the least capacity to adapt or avoid climate hazards altogether and are generally forced to leave as a last resort. Tragically, those even more vulnerable are forced to remain behind under more distressing circumstances.

A likely scenario in the coming decades is that climate-induced emigration will occur where livelihood infrastructure and practices are compromised by climate change. Vulnerable places will include low-lying cities, coastlines vulnerable to sea level rise, and areas of high water and agriculture stress. The Groundswell report suggests that the northern highlands of Ethiopia, the Ethiopian capital Addis Ababa, Dar Es Salaam (Tanzania) and Dhaka (Bangladesh) are likely to be such hotspots of vulnerability.

The other side of the trend sees climate induced immigration where certain places with more favorable climate conditions attract climate migrants due to their ability to more easily sustain agricultural practices and provide other socio-economic opportunities. Again, the Groundswell report identifies some examples, including '...the southern highlands between Bangalore and Chennai in India, the central plateau around Mexico City and Guatemala City, and Nairobi in Kenya.'

Adapting to emerging climate issues

As the impacts of climate change become increasingly apparent across the globe, there is a growing imperative to implement effective adaptation strategies. Communities and ecosystems around the world are faced with unprecedented challenges, such as more intense extreme weather events. Climate change *adaptation* refers to the process of adjusting to these changes in a manner that minimizes risks to the built environment and natural systems and secures the ideals of sustainable development.

Effective adaptation strategies are necessarily multifaceted and context-specific; they need to address a wide range of vulnerabilities across different sectors and regions. In urban areas, for example, adaptation might involve the development of green infrastructure and resilient building systems to manage heatwaves and floods. In the agricultural sector, it could include the use of drought-resistant crops and more efficient irrigation systems to safeguard food security. By tailoring approaches to local needs and conditions, adaptation efforts can significantly mitigate the adverse effects of climate change. Adaptation strategies are vital for building resilience and promoting sustainable development in the face of climate change.

Just as important, it should be noted that climate change adaptation is not just about survival; it is much more than that. It entails the promotion of real opportunities for sustainable growth and development, especially by investing in innovative solutions, strengthening institutional capacities, and fostering community engagement to build more resilient societies. As we navigate the complexities of a changing climate, it is crucial to prioritize adaptation (alongside mitigation efforts) to ensure a holistic and effective response to one of the most pressing global challenges of our time.

Some adaptation solutions to the emerging climate issues discussed are listed below to guide the interested reader in the right direction.

Adaptation Solutions		
Climate Change	Food Security	Climate Migration
Renewable Energy Transition Increasing the use of renewable energy sources like solar and wind energy	**Sustainable Agriculture Practices** Promoting practices which improve soil health and yields while not harming the environment.	**Policy and Legal Frameworks** Create policies that provide legal protection and support for climate migrants
Policy Frameworks National and international policies need to support climate adaptation strategies	**Technological Innovation** Leveraging technology to enhance food production efficiency, e.g., vertical farming	**Resilient Infrastructure** Build climate-resilient infrastructure in vulnerable areas to reduce the need for migration
Climate-Resilient Agriculture Promoting climate-resilient agricultural practices	**Food Storage and Distribution** Improving food storage facilities and distribution networks to reduce post-harvest losses.	**Economic Opportunities** Create economic opportunities in vulnerable regions to reduce the push factors for migration

FACTS

- Every year, around five trillion plastic bags are used worldwide.
- Each year, 400 million tonnes of plastic are manufactured worldwide.
- Only 9% of the plastic ever created has been recycled globally, whereas 79% is currently in landfills, dumps, or the environment, and 12% has been burned.
- The annual trash generation is anticipated to increase by 70% from 2016, to 3.4 billion tonnes in 2050, due to rapid population expansion and urbanization.
- By 2050, the plastics sector will be responsible for 20% of all global oil consumption, if current trends hold.
- 40% of all waste is created during building construction and later demolition.

Growing influence of stakeholders

The growing influence of stakeholders in relation to sustainability and environmental, social, and governance (ESG) issues has emerged as a significant trend in recent years. Stakeholders, including investors, consumers, employees, communities, and regulators, are demanding greater transparency, accountability, and action from organizations on sustainability and ESG matters. This shift is driven by various factors, such as the recognition of businesses' impact on the environment and society, the rise of socially responsible investing, and the evolving regulatory landscape. This section aims to explore the increasing influence of stakeholders in promoting sustainability and ESG practices, drawing upon research papers and reliable resources.

1. INCREASING INVESTOR FOCUS ON SUSTAINABILITY AND ESG

Institutional investors are increasingly integrating ESG factors into their investment decisions, recognizing the potential impact on financial performance and risk management. A study published in the *Harvard Business Review* found that firms with strong ESG performance have a lower cost of capital and higher market valuation, indicating that investors value sustainability and ESG practices. Moreover, research conducted by the Global Sustainable Investment Alliance indicates that sustainable investments have experienced significant growth, with US$30.7 trillion of assets under management employing sustainable strategies in 2018.

2. CONSUMER DEMAND FOR SUSTAINABLE AND RESPONSIBLE PRODUCTS

Consumers are increasingly demanding sustainable and responsible products. According to a global survey by Nielsen, 81% of consumers feel strongly that companies should help improve the environment. Research published in the *Journal of Consumer Research* highlights that consumers are willing to pay a premium for sustainable products and are more likely to choose brands that align with their values. Furthermore, the same study suggests that sustainability can positively influence consumer perceptions of product quality and brand reputation.

3. EMPLOYEE EXPECTATIONS AND TALENT ATTRACTION

Employees, particularly younger generations, are increasingly prioritizing sustainability and expect their organizations to actively address environmental and social issues. A study by Deloitte found that 74% of millennial employees believe that their organization should be more actively involved in addressing sustainability issues. Additionally, organizations that prioritize sustainability and ESG practices are more attractive to job seekers. The Harvard Business Review emphasizes that sustainable organizations are particularly appealing to younger generations who seek purpose-driven work and want to contribute to meaningful causes.

4. COMMUNITY AND REGULATORY PRESSURES

Communities are exerting growing pressure on businesses to adopt sustainable and responsible practices, particularly in industries with significant environmental and social impacts. Stakeholders within communities are becoming more vocal in their demands, urging businesses to minimize their environmental footprint, respect human rights, and contribute positively to local communities. Simultaneously, governments and regulatory bodies are implementing ESG-related regulations and policies. For instance, the European Union's Sustainable Finance Disclosure Regulation (SFDR) requires financial institutions to disclose information on their ESG practices and impacts, influencing organizations to take ESG considerations more seriously.

5. BUSINESS BENEFITS OF SUSTAINABILITY AND ESG

Prioritizing sustainability and ESG practices can yield various benefits for businesses. Research published in the MIT Sloan Management Review suggests that companies with a sustainability focus outperform their peers in terms of operational efficiency, brand reputation, and customer loyalty. These organizations are better positioned to adapt to changing market expectations and regulatory environments, mitigating risks associated with reputation damage and legal liabilities. Additionally, a study published in the Journal of Sustainable Finance & Investment indicates a positive correlation between a firm's sustainability performance and its financial performance, emphasizing the potential for long-term value creation through sustainable practices.

The growing influence of stakeholders regarding sustainability and ESG is reshaping the business landscape. Investors, consumers, employees, communities, and regulators are increasingly demanding transparency, accountability, and action from organizations in these areas. Organizations that proactively address sustainability and ESG issues can gain a competitive advantage, enhance their reputation, attract and retain talent, and better manage risks. By incorporating sustainability and ESG practices into their strategies, organizations can navigate the evolving expectations of stakeholders and contribute to a more sustainable and responsible future.

Environmental, social, & governance criteria

Environmental, social, and governance (ESG) criteria have gained significant attention in recent years, as investors and businesses recognize the importance of sustainable and responsible practices. This section aims to provide a comprehensive analysis of ESG factors by examining their individual components, the impact on financial performance, and the growing adoption of ESG integration. Drawing on research papers and reliable resources, we explore the environmental, social, and governance aspects of ESG criteria and their implications for organisational decision making, particularly within the investment community.

In the face of global challenges such as climate change, social inequality, and corporate scandals, there has been a growing focus on responsible investment and sustainable business practices. ESG criteria have emerged as a framework to assess the sustainability and ethical impact of investments. Key elements of the 3 factors are provided below as an elementary overview:

1. ENVIRONMENTAL FACTORS

Environmental criteria evaluate a company's impact on the natural environment. These impacts include resource consumption, pollution levels, carbon emissions, and commitment to environmental stewardship. Research has demonstrated a positive correlation between strong environmental performance and financial performance. Companies with robust environmental practices tend to have better long-term financial results and reduced costs of capital.

2. SOCIAL FACTORS

Social criteria encompass a company's impact on stakeholders, including employees, customers, communities, and suppliers. When evaluating factors such as labor practices, human rights, diversity and inclusion, and community engagement, social performance has been linked to improved financial performance. Companies with strong social practices experience reduced employee turnover, enhanced employee productivity, and increased customer loyalty.

3. GOVERNANCE FACTORS

Governance criteria focus on a company's internal systems and structures, including its decision-making processes, transparency, and accountability. Research has highlighted a positive relationship between good governance practices and financial performance. Companies with robust governance structures are more likely to attract long-term investors, maintain their reputation, and mitigate risk.

4. INTEGRATION OF ESG FACTORS

Integrating ESG factors into investment strategies has gained traction due to the potential benefits it offers. Research shows that ESG integration can lead to improved risk-adjusted returns and enhanced risk management. By considering ESG factors in investment decision making, investors can identify and assess material risks and opportunities that traditional financial analysis may overlook. The Principles for Responsible Investment (PRI) network conducted a study across asset classes and found that ESG integration improves risk management and decision-making processes.

5. FINANCIAL PERFORMANCE AND ESG FACTORS

Numerous studies have investigated the relationship between ESG factors and financial performance. A meta-analysis of 2,200 studies concluded that there is a positive correlation between corporate social responsibility (CSR) and financial performance. Moreover, research conducted by Morgan Stanley showed that integrating ESG factors into investment strategies can lead to improved risk-adjusted returns.

6. THE GROWING ADOPTION OF ESG CRITERIA

The adoption of ESG criteria has witnessed significant growth in recent years. The Global Sustainable Investment Alliance reported that the global sustainable investment market reached $35.3 trillion in assets under management, reflecting the increasing adoption of ESG criteria. Governments, regulators, and institutional investors are increasingly emphasizing the importance of ESG factors in investment decision making, further driving the integration of ESG considerations.

ESG criteria, encompassing environmental, social, and governance factors, provide a robust framework for assessing the sustainability and ethical impact of investments. They have been associated with improved financial performance, enhanced risk management, and stakeholder relationships. The growing body of research and industry adoption underscores the importance of considering ESG factors in investment decision-making processes. As investors and businesses strive for long-term sustainable growth, ESG criteria offer valuable insights into companies' practices and their potential impact on financial performance and societal well-being.

Suggested models to co-align sustainability and ESG

This section explores suggested models to co-align sustainability and environmental, social, and governance (ESG) considerations. The integration of sustainability and ESG practices has gained significant attention in recent years as businesses recognize the need to create long-term value while addressing social and environmental challenges. Through an analysis of research papers and reliable resources, this study highlights several models and frameworks that can help organizations effectively align sustainability and ESG factors. The identified models include the sustainable development goals (SDGs), the triple bottom line (TBL), integrated reporting, and the ESG materiality framework. The analysis provides insights into the key features, benefits, and limitations of each model, facilitating a comprehensive understanding of their applicability and potential synergies for organizations seeking to integrate sustainability and ESG practices.

In the face of global challenges such as climate change, social inequality, and governance issues, businesses have recognized the importance of integrating sustainability and ESG considerations into their strategies. This integration involves incorporating environmental, social, and governance factors into decision-making processes to create long-term value, mitigate risks, and enhance stakeholder trust. This section aims to explore several suggested models that can help organizations align sustainability and ESG effectively.

SUSTAINABLE DEVELOPMENT GOALS (SDGS)

The sustainable development goals, established by the United Nations in 2015, provide a comprehensive framework for addressing global challenges. The 17 goals cover a wide range of sustainability issues, including poverty eradication, quality education, climate action, and responsible consumption. By aligning their operations, strategies, and reporting with the SDGs, organizations can demonstrate their commitment to sustainable development and contribute to global targets.

The SDGs offer a common language and set of aspirations that enable businesses to identify areas where they can make a positive impact. By mapping their activities and initiatives to specific SDGs, organizations can align their sustainability efforts with global priorities. Additionally, integrating SDGs into business strategies can help identify market opportunities, enhance brand reputation, attract investors, and foster innovation. However, the main challenge lies in selecting relevant SDGs and translating them into actionable targets that align with an organization's unique context and business operations.

TRIPLE BOTTOM LINE (TBL)

The triple bottom line framework, developed by John Elkington, suggests that organizations should measure their performance based on three dimensions: social, environmental, and financial. This model recognizes that economic success alone is not sufficient and encourages organizations to consider their impacts on people and the planet. By adopting the TBL approach, companies can develop a holistic understanding of their value creation, promoting responsible decision-making and stakeholder engagement.

The TBL framework provides a comprehensive view of organizational performance by considering social and environmental impacts alongside financial outcomes. It enables organizations to assess their contributions to society and the environment, encouraging them to go beyond profit maximization. By measuring and reporting on social and environmental factors, businesses can enhance transparency and accountability, building trust with stakeholders. However, challenges arise in accurately quantifying social and environmental impacts and aligning them with financial metrics.

INTEGRATED REPORTING

Integrated reporting is a model that aims to provide a concise and comprehensive overview of an organization's strategy, governance, performance, and prospects, incorporating both financial and non-financial information. This model facilitates the integration of sustainability and ESG factors into organizations' reporting practices, promoting transparency and accountability.

Integrated reporting enables organizations to present a holistic view of their value creation by considering both financial and non-financial factors. By reporting on ESG performance, companies can demonstrate their commitment to sustainable practices, attract responsible investors, and engage stakeholders effectively. Integrated reporting helps organizations assess their long-term value creation and identify areas for improvement, facilitating strategic decision-making and capital allocation. However, challenges exist in identifying and standardizing non-financial indicators and effectively communicating complex information to diverse stakeholders.

ESG MATERIALITY FRAMEWORK

The ESG materiality framework helps organizations identify and prioritize the environmental, social, and governance issues that are most relevant to their business operations and stakeholders. By conducting materiality assessments, companies can determine the key ESG factors that have a significant impact on their financial performance and reputation.

The ESG materiality framework assists organizations in understanding the ESG issues that are most material to their business. By focusing on material ESG factors, companies can allocate resources effectively, set meaningful targets, and develop targeted strategies to address the most significant sustainability challenges. Materiality assessments involve engaging with stakeholders to understand their concerns and expectations, ensuring that organizations address the most relevant issues. Nonetheless, challenges lie in defining materiality thresholds, engaging with stakeholders, and keeping up with evolving sustainability trends.

COMPARATIVE ANALYSIS AND SYNERGIES

By comparing the identified models, it becomes evident that they share common features and objectives. For instance, the SDGs and TBL both emphasize the importance of multiple dimensions of value creation, while integrated reporting and the ESG materiality framework facilitate the integration of sustainability and ESG factors into reporting practices. Organizations can leverage the synergies between these models to develop a comprehensive approach to sustainability and ESG, aligning their strategies, operations, and reporting effectively.

Integrated reporting can incorporate the SDGs into its reporting framework, allowing organizations to demonstrate their contributions to the global goals. Similarly, the TBL framework can be utilized to measure and report on ESG factors, expanding beyond financial metrics. Combining these models can provide a holistic view of an organization's performance, encompassing economic, social, and environmental aspects. The ESG materiality framework can be integrated into both the TBL and integrated reporting to identify the most material ESG issues and ensure targeted actions.

Integrating sustainability and ESG considerations is crucial for organizations to create long-term value and address global challenges. The models discussed in this section, including the sustainable development goals, triple bottom line, integrated reporting, and the ESG materiality framework, offer valuable tools for aligning sustainability and ESG practices. Each model has its own benefits and limitations, and organizations should carefully assess their unique context and objectives when selecting and implementing these models. By leveraging the synergies between the identified models, organizations can develop a comprehensive and robust approach to sustainability and ESG, promoting responsible business practices and creating positive impacts for stakeholders and the planet.

Climate, energy, and environmental sustainability through ESG lenses

Climate, energy, and environmental sustainability have become critical global issues, driving the need for effective strategies and frameworks to address them. One approach that has gained significant traction in recent years is the integration of environmental, social, and governance (ESG) factors into decision-making processes. This section will delve into the topic of climate, energy, and environmental sustainability through the lens of ESG, examining the relationship between ESG and sustainable development, the impact of ESG on climate change mitigation and renewable energy, and the challenges and opportunities associated with ESG implementation.

ESG refers to the set of environmental, social, and governance criteria used to evaluate the sustainability and societal impact of an investment or business. The integration of ESG factors enables investors, companies, and policymakers to assess the long-term risks and opportunities associated with environmental and social issues. By incorporating ESG considerations, decision-makers can drive positive change and sustainable development, aligning financial performance with environmental and social responsibility.

The Paris Agreement, adopted in 2015, marked a turning point in global efforts to combat climate change. ESG principles can play a crucial role in achieving the goals outlined in the agreement. Research by Eccles, Ioannou, and Serafeim (2014) found that high ESG ratings positively correlate with better stock performance and lower cost of capital. This suggests that companies with strong ESG practices are more likely to attract investments and secure financing for climate change mitigation and sustainable energy projects. ESG integration can also drive innovation and technological advancements to accelerate the transition to a low-carbon economy.

ESG factors significantly influence climate change mitigation efforts. A study by Nielsen and Markard (2017) found that ESG criteria can enhance the effectiveness of policies aimed at reducing greenhouse gas emissions. By considering environmental factors, companies can align their operations with climate change mitigation goals, adopt cleaner technologies, reduce emissions, and promote energy efficiency. Furthermore, social factors associated with ESG, such as community engagement and stakeholder involvement, can facilitate the acceptance and adoption of renewable energy projects. For instance, involving local communities in the decision-making process and ensuring equitable distribution of benefits can help overcome resistance and foster social acceptance of renewable energy initiatives.

Renewable energy is a key component of sustainable development, and ESG considerations can drive its adoption. A study by Gantenbein, Volonté, and Wohlwend (2016) found a positive relationship between a firm's environmental responsibility and its investments in renewable energy. Companies with strong ESG practices are more likely to prioritize clean energy investments and diversify their energy sources. Additionally, ESG factors influence the overall sustainability of the renewable energy sector. For instance, governance factors can ensure transparency, accountability, and ethical behavior within the industry. Social factors can ensure fair labor practices and protect the rights of communities affected by renewable energy projects, while environmental factors can safeguard biodiversity and ecosystems.

Despite the potential benefits, there are challenges associated with implementing ESG frameworks. One challenge is the lack of standardized ESG metrics and reporting practices. Research by Clark, Feiner, and Viehs (2015), still valuable a decade after publication and despite the harmonization efforts of recent years by the IFRS sustainability disclosure standards, highlights the need for harmonized standards to enable consistent and comparable ESG reporting. Standardization would facilitate meaningful comparisons among companies and enhance the reliability and transparency of ESG information. Another challenge is greenwashing, where companies falsely portray their environmental and social performance. This highlights the importance of independent verification and rigorous assessment frameworks to ensure the credibility and integrity of ESG ratings.

The rise of ESG has also brought opportunities for innovation and collaboration. The integration of ESG factors encourages the development of new technologies, business models, and investment strategies that align with sustainability goals. For instance, impact investing, which aims to generate positive social and environmental outcomes alongside financial returns, has gained popularity as a result of ESG integration. Collaboration between stakeholders, including investors, companies, governments, and civil society, is crucial for addressing complex sustainability challenges. By working together, stakeholders can leverage their expertise and resources to drive systemic change and create a more sustainable future.

Climate, energy, and environmental sustainability can be effectively addressed through the lens of ESG. ESG integration provides a framework for evaluating the long-term risks and opportunities associated with environmental and social issues, enabling decision makers to drive positive change and align financial performance with sustainability goals. ESG factors influence climate change mitigation efforts, renewable energy adoption, and the overall sustainability of various sectors. However, challenges such as the lack of standardized metrics and greenwashing need to be addressed. Despite these challenges, the rise of ESG brings opportunities for innovation and collaboration, fostering the development of new technologies and investment strategies that promote sustainable development. By embracing ESG principles, we can pave the way for a more sustainable and resilient future.

VI

The Policy Environment

"It's always cozy in here. We're insulated by layers of bureaucracy."

To ward off the effects of climate change, the international community has to take urgent and coordinated action. Although there is virtually universal consensus on the need to act, there are significant policy challenges in adopting universal mitigation and adaptation measures to combat the problem. To complicate matters, the responsibility for responding, and capacity to respond, to climate change are unequally distributed. Indeed, the historical responsibility for climate change is one of the sticking points of climate change negotiations. By some accounts, developed countries are responsible for nearly 80% of historical carbon emissions. To complicate matters even further, the profile of emissions has changed in recent decades, and in the first decade of the 21st century, developing countries overtook industrialized countries in the generation of these emissions (WRI, 2024).

A great majority of GHG emissions come from a small number of nations. These nations also happen to be the same parties that have the technological and financial resources to provide many of the required solutions. Power inequality is therefore inextricably linked to climate change and the development of policy solutions.

The international community recognizes the need for action on climate change. But given the ponderous bureaucratic mechanisms of multilateral processes and the geopolitical challenges of 'north' vs. 'south' environmental politics, it is unclear whether a timely and effective response can emerge.

A cursory assessment of the history of environmental politics and climate negotiations suggests that a truly effective and global solution may not be imminently at hand. In climate negotiations, it is critical to iron out political conflicts between powerful blocs as a first step toward actual cooperation.

The outcomes of climate negotiations are determined to a large degree by the formation and dynamics of groupings. While the UNFCCC categorizes countries into five groups (African states; Asian states; Eastern Europe; Latin America and the Caribbean; and Western Europe and 'Other States, which include Australia, Canada, Iceland, New Zealand, Norway, Switzerland, and the United States), it also divides countries in three categories: developed countries, developed countries with special financial responsibilities, and developing nations.

However, a number of informal and formal blocs have emerged over the years. These blocs engage in 'diplomatic wrestling' to push common interests. Even though climate negotiations were, for a long time, perceived as too important to be victims of global politicking, that is no longer the case and, in fact, climate negotiations are a highly political experience. Some countries even operate in multiple blocs, while it is also often the case that countries clustered together may have different views on climate issues.

The most common blocs include:

- European Union (EU): The EU bloc represents the nearly 30 member states who typically agree on a single negotiating position.
- G77 and China: This bloc has become larger and increasingly diverse, with its members (over 130) often holding diverging opinions. But it is also capable of holding a united front and has, in recent times, put up a united front, especially on issues like financial support.
- Least Developed Countries (LDCs): Nearly 50 of the world's less developed economies are represented in this bloc. It typically advances views on capacity building to enhance ability of member states to manage climate change impacts. The LDC bloc is also part of the larger G77 and China bloc, although it may sometimes hold different views from the emerging economies.
- Small Island Developing States (SIDS): This bloc of many low-lying islands vulnerable to sea-level rise often presents a united front during negotiations and is quite bullish in its quest to ensure reductions on GHG emissions.
- Alliance of Small Island States (AOSIS): This bloc of over 40 low-lying and small island countries, most of which are members of the G77, is united by the common vulnerability of its members to sea-level rise. It functions primarily as an ad hoc lobby and negotiating voice for Small Island Developing States (SIDS).
- The Umbrella Group: A group of mostly developed countries with some particularly high-emitting countries like the US, Australia, and Canada. At its core, this bloc holds that GHG reductions required as part of the Paris Agreement must come from all countries, including developing countries. In essence, it prefers that climate

responsibility be based on current rather than historical emissions, and is therefore ideologically in opposition with a bloc like the LDC.

- BASIC: This bloc consists of the emerging economies of Brazil, South Africa, India and China, which came together after COP15 Copenhagen having felt increasing pressure to control their emissions.

There are other blocs that work together in the name of climate negotiations (e.g., the Coalition for Rainforest Nations, and the Climate Vulnerable Forum). The dynamics of climate negotiations are influenced most especially by the blocs discussed above. During negotiations, blocs will jointly support certain policies or viewpoints. This makes climate negotiations an extremely political process and one which must delicately balance scientific and political views. Getting this balance right is an extremely challenging process and requires the most astute skills in negotiation, technical management, and procedural administration. If this balance is not achieved, years of negotiations can be compromised at a single conference.

To illustrate the difficulty of achieving the right balance at climate negotiations, let us look at some examples of significant climate conferences over the years.

COP6, The Hague (2000)

In November 2000, the Hague played host to the sixth session of the United Nations Framework Convention on Climate Change (UNFCCC) Conference of the Parties (COP6).

The key objectives of COP6 were to agree on the implementation details of the Kyoto Protocol to make it ratifiable for early entry into force. It was intended to determine the specific rules to make the Kyoto Protocol commitments more clearly operational in a way that would enable measurement of the emissions reductions. Although it was a historic agreement, the Kyoto Protocol contained a lot of vague language on emissions reduction measures. Unfortunately, the ambiguity of this language was a factor that would contribute to some of the challenges of COP6. Additionally, the Kyoto Protocol was agreed at a time of great expectations, and as the EU in particular forced the agenda for real emissions reductions, the negotiation process extracted a price for such binding commitments in the form of inclusion of carbon sinks. As we shall see, this inclusion became a particular sticking point during COP6.

As COP6 attempted to piece together a plan for the design and implementation of operational details of Kyoto, it became clear that some discussion points raised a significant divergence of opinions. The key issues that would need significant attention included carbon sinks, 'hot air' and the role and capacity of institutions among others.

The issue of carbon sinks turned out to be a particularly contentious one. The core issue that needed to be resolved was what amount and proportion of overall emissions reductions should countries receive as credits for forestry and land-use activities. The sticking point was the uncertainty over how much carbon soils and green matter actually remove from the atmosphere. The use of carbon sinks as an aid to emissions reductions was therefore not only a politically charged issue but also a technically complex one.

The stage was set for negotiations, which were primarily attended by the Umbrella Group (US, Canada, Australia, New Zealand, Japan), the EU bloc, G77 and China, and the Alliance of Small Island States (AOSIS). It was virtually impossible for these parties to reach agreement on some key issues, particularly carbon sinks and other points, like financing.

A particularly vexatious negotiation topic was the issue of sinks (activities in the land-use, land-use change, and forestry – LULUCF – sector that remove greenhouse gases from the atmosphere) under the Kyoto Protocol. The Protocol allowed Annex I parties (developed economies) to use sinks to offset their emissions targets, but there was no final agreement on when and to what extent sinks could count toward emissions reductions commitments. Some parties, like the US, were interested in using sinks to achieve their commitments under the Kyoto Protocol; in fact, the Umbrella Group considered inclusion of sinks as a precondition for their ratification of the Protocol. On the other side of the negotiations, the EU worried that countries would use sinks to account for a majority of their emissions-reductions commitments, and was particularly keen to avoid the inclusion of soils as carbon sinks.

Other areas of discussion were the separation of human and natural effects, and the possibility of including sinks within the CDM (one of the 'flexibility mechanisms' to facilitate emissions reductions under the Kyoto

Protocol). Even under these areas there was considerable debate. For instance, on the question of whether to allow sinks in the CDM, the Umbrella Group demanded crediting of their carbon sinks within the CDM, while the EU wanted to completely exclude sinks from the CDM at least during the first commitment period (2008–2012). Additionally, there were fractures in positions from within the European parties. Countries like Finland and Sweden began to consider parting from the EU position as they contemplated the potential gains of their domestic deforestation activities. It is clear that the EU had underestimated the importance of carbon sinks during COP6.

Ultimately, the talks collapsed as blocs held to their positions firmly. While there was a desire to push for ratification of the Kyoto Protocol, there was also a need to maintain the environmental integrity of the Protocol. There were significant recriminations among the negotiation parties. The EU accused the Umbrella Group of trying to disable its Kyoto targets through constant demands for loophole arrangements. On the other hand, the US accused the EU of being inflexible and arrogant, and even attempting to subvert the industrial bases of other leading economies. At the very moment that clarity of positions was required, internal differences revealed themselves within the EU bloc. From a procedural perspective, and given the legacy of ambitions set in Kyoto without agreement on finer operational details, COP6 was also hampered by a short negotiation period.

COP15, Copenhagen (2009)

The events of COP15 in Copenhagen, sometimes described as 'Hopenhagen' in the lead-up period, also amply demonstrate some of the challenges of climate negotiations. The participating countries attended the conference with the objective of piecing together a comprehensive global climate agreement to replace the Kyoto Protocol which, although ratified in 2005, only obliged GHG reductions from a few developed economies.

Leading to the conference, the European Commission outlined its aims for a deal it hoped would be endorsed globally. These aims included targets and actions, climate finance for low-carbon initiatives and adaptation, and the development and harmonization of a global carbon market. The Europeans made clear their preference for such aims to be legally binding on parties in order to compel real emissions reductions.

Hopes were therefore high that the conference would establish a successor agreement to the Kyoto Protocol, which was due to expire in 2012. Unfortunately, things fell apart at the conference primarily due to procedural shortcomings. First of all, negotiators were not prepared for the eventual arrival of politicians, because the Danish Prime Minister's team replaced the pre-conference negotiators who had already built significant goodwill with various parties. This disconnect led to the leaking of proposed measures to maintain average global temperature rise to 2°C above pre-industrial levels within an agreement presumptuously labelled the Copenhagen Agreement. This so-called 'Danish text' infuriated many negotiating parties at the conference, most of which were developing economies, which claimed they were simply being excluded from the negotiation process. China and India were particularly forceful in their displeasure about what they deemed as developed economies designing an agreement that suited themselves without seeking the views of developing economies. Additionally, from a procedural standpoint, the Danish government selected a venue that was too small to host the number of attending delegates. This became an acute problem after the first week of the conference, when leading politicians arrived. Consequently, other participants, such as NGOs and media observers, were denied entry to the conference. The third, and perhaps most significant, procedural issue related to the actual agreement itself. Toward the last day of the conference, it was dawning on observers that the conference was unlikely to be a success.

The UN climate negotiation process works by consensus, and every country can express disagreement; many countries did express their disapproval of the proposed deal. Just as the conference seemed headed for complete failure, President Barack Obama arrived in Copenhagen on the scheduled final day of the conference. He held a series of meetings with Chinese Premier Wen Jiabao, Indian Prime Minister Manmohan Singh, Brazilian President Luiz Inacio Lula da Silva, and South African President Jacob Zuma, and together they drafted a document which became the Copenhagen Accord. This unplanned assembly of leaders to draft a deal was very unusual, since heads of state do not usually participate in actual negotiations, much less lead them. These 'extra UN' negotiations culminated with Obama declaring a victory with a statement of outcomes agreed upon with the four other leaders. Ultimately, the outcomes of the Copenhagen Accord were presented to the 180+ remaining UN member states as a done deal, with some accepting it and others denouncing it.

Although the accord featured some key elements, such as an acknowledgment of the need to hold temperature increases to below 2°C, and targets by developed economies to limit GHG emissions by 2020, it was non-binding. Additionally, it became clear that any Copenhagen Accord signatories were effectively stepping out of the Kyoto Protocol and the UN process. Ultimately, countries conceded to this unusual approach and adopted parallel decisions under the UNFCCC and the Kyoto Protocol to simply 'take note' of the Copenhagen Accord and allow governments to individually sign on.

Dynamics of climate policy negotiations

Climate policy negotiations are structured around coalitions of blocs, which are groups of countries that band together around common interests. There are differences in how the various blocs approach negotiations; not only that, the way that blocs have operated since the earliest climate policy negotiations of the early 1990s has changed. The ratification and eventual succession of the Kyoto Protocol in particular spurred the creation of numerous blocs, as countries and regions sought ways to address a growing number of issues and commitments.

Climate negotiations are complex affairs, and blocs reduce this complexity by increasing the ability and impact of individual countries to negotiate, and also by reducing the number of actors in the negotiation process. Nations are compelled to cooperate in blocs, not just because of shared interests on issues; they must also have shared objectives.

However, blocs are, in and of themselves, a significant factor in the negotiation process. There are significant administrative, procedural, and political efforts required to maintain them, and this fact alone makes them a particular challenge for developing economies. Additionally, the nature of coalitions necessarily entails some level of compromise on positions, and feedback with source national authorities to endorse country representation on bloc positions. Consequently, bloc positions do not necessarily equal individual country positions, and often reflect a degree of compromise on these positions.

Despite this element of compromise, blocs still represent good value for countries. Indeed, for the less influential countries, blocs are the only way to negotiate meaningfully with global leading powers like the United States. It is not a surprise that leading powers hold a dominant position in the multilateral world of climate negotiations. They tend to have significant economic capacity, technological advancement, and institutional maturity, and as a result, their influence on global climate issues is particularly powerful. Coupled with their historical industrial bases, they are particularly important players in climate change solutions. In the face of such power, less influential countries see the bloc as a better position from which to attempt influence of their own.

Broadly speaking therefore, the bargaining agenda in global climate negotiations tends to be highlighted by issues split along the developing country and developed country line. This is perhaps the most common fault line which defines the complexities of climate negotiations.

The dynamics of climate policy negotiations are defined by a number of characteristics, among them the specificity of issues, the persistence of blocs, the hierarchy of blocs and the overlapping of bloc memberships (Weiler, Castro, and Klöck, 2021).

Specificity of issues

Ever since we have organized ourselves as independent political communities, humans have experienced an uneven distribution of power, knowledge, and competences. Generally speaking, countries with greater power, knowledge, and competences can exercise their interests in ideology more comfortably than other members of the international community. In this context, greater powers have been able to assert their industrial might over the years and have left a larger footprint of these activities. It is within this setting of different actors in the global international scene that climate change discussions have taken place over the years. Of the actors, it is noteworthy to mention that 'great powers' have used their capabilities to run the agenda of special issues like climate change. Great powers can either promote environmentally deleterious actions, or they can be champions of positive initiatives on a global scale. Either way, they wield environmental power on a regional and global scale. The willingness to take responsibility for this power is an important factor in the engagement of state actors in global environmental politics.

All these elements – international actors, great powers, positive and negative actions, and the responsibility of power – play a role in determining the specificity of issues that are topical at any one time in global environmental policy discussions. Blocs emerge as a result of the interplay of these elements. Blocs for addressing climate change were established at specific times when issues related to climate change were topical at the negotiation table.

We have already discussed the debacle at COP15 in Copenhagen, where parties failed to secure a legally binding global treaty to reduce carbon emissions. The Cartagena Group/Dialogue for Progressive Action was set up as an informal grouping of countries to reduce the rancor between developed and developing economies, such as Australia, Norway, Kenya, and Indonesia. The Cartagena Group took a proactive and pragmatic approach to promote the objective of reaching a comprehensive and legally binding global agreement. The group allowed countries to explore positions beyond their stated formal positions without risk of national reprimands or media scrutiny. In other words, the Cartagena Group emerged specifically because the stiff negotiating environment of the UNFCCC process had stifled negotiation progress up to that point, and the Cartagena Group seemed to provide a safe environment for negotiating parties to seek areas of convergence, which they could still take back to their own formal blocs.

PERSISTENCE OF BLOCS

Once formed, blocs tend to remain active even beyond their original agendas. One might wonder why blocs remain even after the attainment or abandonment of their original agendas; there are some compelling reasons. The financial and administrative demands of bloc formation and management mean that members are often cautious about disbanding the blocs. The creation of an extra voice is often valid for the sake of the platform it provides and countries do not want to relinquish that platform. A good example has been presented by eminent authors who discuss the case of the Like-Minded Developing Countries (LMDC) bloc (Weiler, Castro, and Klöck, 2021). The LMDC is a bloc of dozens of countries which is one of the most influential groups of developing-country parties in climate policy negotiations, and includes powerful parties such as China, India, and Saudi Arabia.

In the words of former Bolivian Minister of Development Planning Rene Orellana Halkyer, the LMDC views itself as 'the representative of the voices that call for a more equitable world.' This is true in the sense that the LMDC aims to uphold the dual recognition of developed and developing economies as strictly as possible, with particular emphasis on binding greenhouse gas emissions reduction targets only for developed economies. However, the Paris Agreement requires all signatories, whether developed or developing economies, to commit to nationally determined contributions to the overall target of reducing emissions. As a result, the original purpose of the LMDC has already been overtaken by events. However, the LMDC has continued to find relevance and mobilize resources and coordinate actions as required. For instance, signatories to the Paris Agreement agreed to monitor and periodically review their collective progress toward meeting the long-term temperature goal (known as the global stock take [GST]). The LMDC has pushed for specific actions on the GST, such as advocating for UNFCCC's principles of equity and common but differentiated responsibilities (CBDR) *based on historical emissions*. It also typically objects to the singling out of particular sources of emissions (like fossil fuels) or references to the 1.5°C temperature goal and associated mitigation actions.

It is reasonable to assume that the activity level of blocs fluctuates primarily to the tune of specific agenda items. There are times when blocs are particularly vocal and more openly engaged on issues, and other times where they seemingly exist only in theory.

Two legs good, four legs better

Within the context of climate policy negotiations, there is a hierarchy of blocs. It is also useful to point out that blocs come in broadly two types, characterized by the degree of their formalities. At one end of the spectrum, 'formal' blocs have official member lists and formally advocate the consolidated view of members on issues of consensus. Decisions on issues are usually determined by formal, consensus-driven processes. Such blocs may be governed by a formal constitution and elected office holders. Examples of formal blocs are the African Group, AOSIS, the European Union, and LDCs. At the other end of the spectrum are the 'informal' blocs, which involve a less rigid membership process, less structured governance, and simply provide a platform for the dissemination of information, as opposed to the active prosecution of unified views on issues. Examples of these informal blocs include the Cartagena Group, the Umbrella Group, and the EIG. Where members of informal blocs may promote their own national interests during climate negotiations, the national views of members of formal blocs must at least be aligned in principle with the bloc view.

The most basic of bloc activities were the earliest activities of climate negotiations which spawned blocs such as AOSIS. These earliest blocs tended to be more formal, and characterized by large memberships, global spans, and defined by themes. The thematic focus tends to draw them actively into a wide range of issues beyond their original mandate.

Although the formal nature of their organization enables a more cohesive approach on issues, this cohesiveness is offset by the large size and global span of the blocs. In fact, as the profile of climate change policy has grown over the years, the scale of issues has resulted in internal differences of interests within blocs. As these divergences have proliferated, sub-blocs have formed within the blocs to address the more specific interests within blocs.

Original blocs, therefore, exist side by side, with sub-blocs under their umbrella, both addressing various sides of the issues. This arrangement provides flexibility for membership of blocs to be perpetuated despite differences within their membership ranks. Sub-blocs also allow various fragmented units to advocate their views within blocs, and not just the overarching discussions. For example, the Pacific Small Island Developing States (Pacific SIDS) was established to push Pacific views within AOSIS.

Looking ahead

Climate change is often described as the biggest challenge of our times, and a challenge which requires collective action of the global community. The most central channel for such global cooperation has turned out to be through United Nations negotiations and institutions. Perhaps the most significant outcomes of the climate negotiations era have been the Kyoto Protocol and the Paris Agreement. The Kyoto Protocol was adopted in 1997 and entered into force in 2005; it imposed emissions reductions obligations on industrialized economies. In contrast, the Paris Agreement – adopted in 2015 and entered into force in 2016 – currently commits signatories, which comprise virtually all countries of the world, to emissions reductions to keep global temperature increase to well below 2°C.

There is broad consensus on what sort of future impacts to expect, should predicted climate change materialize, including increase in global temperatures, sea level rise, and shrinking glacial cover. Most impacts are likely to arise from variability of weather and increased frequency and intensity of extreme weather events. Increased global warming is likely to lead to long-term changes in the climate system, with repercussions for ecosystems and human communities. Repercussions include likely changes to water resources, impacts on agricultural systems and corresponding food supply, impacts to human health, and, of course, significant impacts on ecosystems and organisms.

It is clear that climate policy negotiations are a necessary step for the global community to take effective action to combat climate change. But this forum has unearthed significant issues among global participants and is itself hampered by some administrative obstacles.

To achieve the ultimate objective of the UNFCCC to stabilize greenhouse gas (GHG) concentrations 'at a level that would prevent dangerous anthropogenic (human-induced) interference with the climate system,' the global community needs to achieve successful outcomes in at least the following fundamental areas:

- Mitigation of climate change: Climate change mitigation refers to initiatives and various efforts to reduce of the net release of human-induced greenhouse gas emissions, which are warming the planet. The scale of reduction efforts is captured by the views of the IPCC's Sixth Assessment Report (6AR) which states that GHG emissions will have to drop by 21% between 2019 and 2030 to stay within the 2°C increase limit of the Paris Agreement (IPCC, 2023). However, it is also clear that the increasing commitments of reduction efforts around the world will be inadequate to meet the Paris targets.

- Adaptation to climate change: The UNFCCC describes climate change adaption as the adjustments in ecological, social, or economic systems in response to real or anticipated climatic stimuli and their effects or impacts. As climate change is generally expected to result in climatic shifts (in temperature, storm frequency, etc.), leading to negative consequences on human societies and the natural environment, adaptation seeks to increase the ability of our environmental, social, and economic systems to cope with a changing climate.

- Building capacity: The capacity of nations to reach rapid mitigation goals in line with the Paris Agreement and ensure effective adaptation strategies is already a key concern. It is fair to state that a majority of nations are unable to adequately build the necessary capabilities to tackle climate change. Managing the effects of climate change will require significant institutional, organizational, and individual capacity building to address the uncertainty of long-term climate impacts, the urgency of actions required across built and natural environments, and the dynamics of public expenditure management and deployment of resources to implement such actions.

- Climate finance: The UNFCCC refers to climate finance as the local, national, or transnational financing – drawn from public, private, and alternative sources of financing – which seeks to support mitigation and adaptation actions that will address climate change. Mitigation and adaptation projects to address climate change require a flow of funds, and this capital can be sourced from both private and public bodies, including national governments, international organizations, and private businesses. From a climate policy perspective, climate finance encompasses the diplomatic commitments for the flow of funds from developed economies to developing economies. In fact, these commitments are codified in Article 9 of the Paris Agreement which states that 'developed country Parties shall provide financial resources to assist developing country Parties with respect to both mitigation and adaptation...'

International cooperation on climate policy faces a number of significant challenges to ensure the successful implementation of these fundamental requirements. These challenges are probably best encapsulated in the principle of the 'global public good' (GPG) concept. Public goods are goods with benefits or costs that are available to all countries, regions, and population groups ('nonexcludable'), and which can be enjoyed repeatedly by any party without compromising the benefits enjoyed by others ('nonrival'). While some goods may be national in nature, such as police protection, the case of greenhouse gas emissions has a global span. The main problem with managing the global nature of GHG emissions (the essence of the climate change problem) as a public good is the difficulty of ensuring collective action without a central authority to finance and provide the public good, as would be the case for a national good such as, say, police protection. Participating parties in climate policy are aware of the inherent advantage in doing as little as reasonably possible while other parties make efforts to reduce their own emissions at their own costs. The resulting ambiguity in the optimal level of effort each party should put has resulted in a climate policy process that is largely driven by special interests and geopolitical concerns.

The issue of climate change as a global public good carries a number of policy challenges. The clearest challenges are the difficulty of controlling the willingness of parties to take advantage of this open resource; the burden of free riders enjoying the benefits of climate action taken by others; and even the sense of apathy that can overcome many stakeholders. Possible solutions may come in the form of resource privatization or government control. To varying degrees, these are the paths that are already being taken. But perhaps the most effective policy response in the long run will be a form of community intervention and monitoring. The more direct engagement of communities in climate change action may be the game changer that catalyzes the outcomes of current climate policy negotiations. Without the more active solicitation of communities around the world, the best efforts of the policy process are unlikely to achieve their aims. And the world will be a poorer environment for it.

A summary of COPs through the years

This table summarizes the annual Conference of the Parties (COP) international climate negotiation meetings that have been taking place under United Nations auspices for the last 20-plus years. The COP is the supreme decision-making body of the United Nations Framework Convention on Climate Change (UNFCCC) and comprises all countries that are, in legal terms, party to the Convention. The COP takes place annually and provides an opportunity for parties to negotiate new measures and review their progress against the overarching objective of the UNFCCC to limit climate change.[3]

COP Conference	Highlights
COP1 Berlin, 1995	The first Conference of the Parties (COP1) took place in Berlin and was presided over by the environment minister of Germany, Angela Merkel. Delegates met to discuss the UNFCCC mandate one year after its establishment and agreed that commitments in the UNFCCC were 'inadequate' for meeting convention objectives, agreeing to reduce GHG emissions beyond the year 2000 through quantitative objectives and specific deadlines.
COP2 Geneva, 1996	COP2 met in Geneva in July 1996 and concluded by noting the 'Geneva Declaration,' which endorsed the IPCC conclusions and called for legally binding objectives and significant reductions in GHG emissions. During this COP, most industrialized economies also submitted their first national GHG inventory.

3 More details about COP are available on the UN Climate Change website: https://unfccc.int/process/bodies/supreme-bodies/conference-of-the-parties-cop.

COP Conference	Highlights
COP3 Kyoto, 1997	This was a particularly important COP, whose aim was to establish a binding protocol of GHG emission reductions. It was significant because it operationalized the UNFCCC by committing industrialized economies (and economies in transition) to approved legally binding targets for reducing GHG emissions. The conference provided innovative, market-based tools for achieving those reductions. Although it was adopted in 1997, the complexities of the ratification process meant that it entered into force in February 2005. The 'Kyoto Protocol' invoked the principle of 'common but differentiated responsibility and respective capabilities' and placed a heavier burden on industrialized economies in recognition of responsibilities for current levels of GHG emissions in the atmosphere. The protocol mainly focused on mitigation, although it also launched the Adaptation Fund to finance adaptation projects in developing countries.
COP4 Buenos Aires, 1998	After the Kyoto Protocol it was clear that many issues would be challenging to resolve. As a result, at COP4, the parties decided to adopt the Buenos Aires Action Plan, which established commitments to complete work on the different Kyoto market-based tools (Joint Implementation, Emissions Trading, and the Clean Development Mechanism), compliance issues, and policies and measures. These topics would remain the focus over the coming decade.

COP Conference	Highlights
COP5 **Bonn, 1999**	Since COP4 was able to at least establish many of the rules and procedures to enact the Kyoto Protocol, at COP5 the parties continued to negotiate aspects of the Buenos Aires agenda to refine technical matters. This conference also marked significant momentum of participation from non-industrialized economies (so-called developing countries), as highlighted by the outgoing COP4 president, who observed that developing countries were quickly becoming a significant source of additional GHG emissions. Also, Kazakhstan formally requested inclusion into the Annex I listing of the Convention (industrialized economies) in order to be bound by binding climate targets. Another non-Annex I economy, Argentina, announced its own emissions target.
COP6 **The Hague, 2000**	As details to the enactment of the Kyoto Protocol approached their conclusion, the enormity of climate politics became apparent for all to see in COP6. The purpose of COP6 was to agree on the details to implement the Kyoto Protocol and make it ratifiable to bring its entry into force. However, the talks collapsed and the conference president even officially 'suspended' COP6. The major issues revolved around the degree of flexibility a country should be allowed in measuring its emissions reductions, what enforcement mechanisms should be in place, and what sanctions, if any, could be applied. Big differences of opinion divided the US, EU, and the Group of 77 blocs, with the US pushing for a more liberal interpretation of the enforcement mechanisms compared to the EU's preferred rigid approach.

COP Conference	Highlights
COP7 Marrakech, 2001	COP7 set the rules for implementing the more detailed provisions of the Kyoto Protocol through the Marrakech Accords. It set rules for establishing flexible mechanisms, such as a GHG emissions trading system, and implementing and monitoring the Clean Development Mechanism (CDM). The conference also agreed that there would be penalties for parties that failed to comply with their reduction obligations. It was also possibly the first COP to result in significant outcomes on climate adaptation, as it made some significant decisions on adaptation, such as the establishment of three funds to support climate adaptation efforts. These developments took place against a backdrop of US President Bush declaring his opposition to the Kyoto Protocol, and, as a result, inadvertently sending a signal on the power of multilateral climate diplomacy.
COP8 New Delhi, 2002	While COP7 marked the conclusion of three years of negotiations on the operational details of the Kyoto Protocol, COP8 was a transition meeting in a sense. The big issue was the matter of protocol ratification and waiting for the Kyoto Protocol to enter into force; would COP8 expedite the complex ratification process and bring the protocol into force? COP8 was dominated by tensions between industrialized and non-industrialized economies over responsibility for addressing climate change. Ultimately, the conference focused on the development needs of the weakest economies by reaffirming that economic and social development and poverty eradication were the first and overriding priorities of developing country parties.

COP Conference	Highlights
COP9 **Milan, 2003**	As countries waited for the Kyoto Protocol to come into force on the back of a complex process, there was disappointment over the failure of Russia to ratify the protocol, further highlighting the growing politicization and polarization of climate politics. However, elsewhere there was some progress, especially in technical aspects such as the modalities regarding afforestation and reforestation (carbon sinks) project activities under the CDM. Ultimately, between COP8 and COP9, the CDM became operational and therefore projects could be kickstarted by 2004.
COP10 **Buenos** **Aires, 2004**	The Kyoto Protocol was still not ratified by enough industrialized economies and was therefore not yet in force. COP10 was significant for its approval of a package of measures on adaptation to climate change and mitigation measures. The agreed program of works was aimed at developing methodologies to deal with adverse effects of climate change, including greater scientific assessment and support for national adaptation programs in developing economies, while also ensuring the integrity of 'clean' projects in these countries. The parties also began looking at the post-Kyoto mechanism, and how to design emissions reduction obligations after 2012, when the first Kyoto commitment period (2008–2012) was scheduled to end.

COP Conference	Highlights
COP11 **Montreal,** **2005**	By the time COP11 took place at the end of 2005, the Kyoto Protocol had finally come into force. The parties could now adopt decisions of the Marrakesh Accords on many details of the flexible mechanisms, land use, land-use change and forestry (LULUCF), and support for developing economies through capacity building and technology transfer, as well as establishing a special climate change fund. Climate risks also featured heavily in the minds of delegates, especially given the particularly extreme weather of 2005. A work program on impacts, vulnerability, and adaptation was created to enhance the capacity of developing economies to better understand and measure vulnerability and adaptation.
COP12 **Nairobi, 2006**	The Nairobi conference zeroed in on a series of decisions, and new initiatives were adopted to support developing countries, those most vulnerable to climate change. Specifically, COP12 focused on four issues: advancing adaptation issues; improving equity and accessibility of the CDM; reviewing mechanisms on technology transfer; and examining what the post-Kyoto commitment period would look like. As adaptation issues were growing in profile, it is worth noting that the adaptation topic was one of the most highly politicized on the basis of accountability, whereby for industrialized economies discussions on adaptation were tantamount to accepting responsibility for causing climate change.

COP Conference	Highlights
COP13 Bali, 2007	The main objective of COP13 was to define a post-2012 global climate regime, i.e., the end of the Kyoto Protocol's first commitment period. The adoption of the Bali Road Map marked the introduction of GHG emissions reduction objectives for developed economies, accompanied by reduction commitments for developing economies, although specific reduction targets were not yet set. The conference also marked significant geopolitical shifts, as emerging economies such as China, India, South Africa, and Brazil accepted the responsibility to reduce their emissions. Aside from the spotlight on these emerging economies, the conference also highlighted the diminishing capacity of the US to influence the global multilateral climate outcomes, despite its global power status. The Bali conference (COP13) prepared the road map for the new protocol, to solve the global warming problem.
COP14 Poznań, 2008	COP14 marked the midpoint of the adoption of the Bali Action Plan in 2007, where the international community agreed to work toward a comprehensive climate agreement by 2009 and the COP15 when the new agreement was expected to be adopted. COP14 also considered further reductions commitments for industrialized economies for the second commitment period of the Kyoto Protocol. While technology transfer was expected to be a key theme of COP14, it was ultimately overshadowed by the work toward the new climate agreement scheduled for the following year. Nonetheless, the conference reached conclusions on the strategic program of the Global Environment Fund (GEP) in regard to development and technology transfer (the so-called Poznań Strategic Program on Technology Transfer).

COP Conference	Highlights
COP15 Copenhagen, 2009	COP15 was the largest gathering up to that point, as countries aimed for an agreement to replace the Kyoto Protocol, due to expire in 2012. Nations agreed on a new climate deal which called for emissions reductions pledges by all major economies, including China and other major developing countries for the first time. However, there was no clear path toward a treaty with binding commitments. COP15 was dubiously noted for the level of political attention it drew, as, by its conclusion, over 100 heads of state and government were in attendance. Key elements of the highly politicized deal included an aspirational goal of limiting global temperature increase to 2°C; broad terms for the reporting and verification of national actions; and a goal for mobilizing $100 billion a year in public and private finance by 2020; and calls for the establishment of a new Green Climate Fund.

COP Conference	Highlights
COP16 Cancún, 2010	If COP15 was marked by drama, COP16 was notable for the sense of accord that governments actively sought as they put aside festering issues. After the highly politicized drama of the previous year, COP16 saw negotiations play out primarily in plenary and informal consultations. This multilateral focus yielded results, weak though they were, including providing for action to reduce emissions by both developing and developed countries, and steps to strengthen finance, transparency, and fundamental aspects of the multilateral climate framework. In essence, COP16 formally adopted key elements of the Copenhagen Accord, including mitigation pledges and a new Green Climate Fund for developing countries (heralded as one of the key achievements of COP16). To be sure, there were significant issues that remained unresolved. For instance, developing economies expressed frustration at the largely unrealized payments of an annual $100 billion pledge by developed economies, and at the EU's insistence on branding these monies as loans, rather than grants. Although COP16 did not deliver a comprehensive and binding post-2012 agreement that negotiators had been targeting since 2007's COP13 (Bali), the agreement nonetheless marked an important step toward supporting action on climate change, particularly in developing economies.

COP Conference	Highlights
COP17 Durban, 2011	COP17 was notable for launching a new process to establish a legal instrument under the UNFCCC which would be applicable to virtually all countries. For years the EU had been seeking a rigid framework to draw countries into climate action and finally it had been able to secure a roadmap of such magnitude, dubbed the Durban Platform for Enhanced Action. This agreement was important in that it marked a stark departure from the contrasting obligations of developed versus developing countries under the Kyoto Protocol; indeed, at its heart was the common applicability of obligations to virtually all countries. Additionally, the conference formally established the Green Climate Fund in support of mitigation and adaptation in developing countries. It also called for stronger actions on the reporting and review of national mitigation efforts.
COP18 Doha, 2012	The 'Doha Climate Gateway' adopted the second commitment period of the Kyoto Protocol via a round of binding GHG emission targets for Australia, Europe, and a number of other developed economies. It also established management milestone in preparation for a 2015 agreement (note the Durban Platform for Enhanced Action from the previous year). The conference yielded modest outcomes in terms of emissions reduction and financing; for instance, the agreement noted the scientific need for limiting the increase in global temperatures 'below 2°C or 1.5°C above pre-industrial levels,' but did not specify how this could be reached. However, the conference achieved what was expected and kept negotiations toward a new climate deal on track.

COP Conference	Highlights
COP19 **Warsaw, 2013**	Negotiators managed a modest package of decisions that kept climate negotiations on track toward a new global agreement. The conference closed with a deal on a loss and damage mechanism to help developing economies cope with climate change impacts. The deal committed countries to a loss and damage mechanism (the Warsaw International Mechanism for Loss and Damage) and committed developed economies from 2014 to providing expertise and aid to countries that suffer from climate-related impacts. Despite the deal, concerns were raised about the lack of detail on the funding, as well as the absence of any obligations on compensation that many developing economies had long been advocating. The conference was also significant for establishing a lenient framework for parties to set their 'intended nationally determined contributions' without presuming the design of what was shaping as the upcoming Paris accord. In essence, a foundation had been set for an ambitious new treaty in 2015.
COP20 **Lima, 2014**	COP20 started with a dynamic sense of initiative, with announcements such as the joint statement by the US and China of their post-2020 emission targets and the announcement of nearly $10 billion in pledges to the new Green Climate Fund. The sense of jubilance quickly gave way to the dread of procedural entanglements that nations knew awaited them, especially on the divisive matter of differentiation between developed and developing economies. In due time the conference adopted the Lima Call for Climate Action, which advanced the design of the much-anticipated Paris Agreement by clarifying key pieces of draft negotiating text.

COP Conference	Highlights
COP21 Paris, 2015	At COP21, parties reached a historic agreement to combat climate change. The core objective of the Paris Agreement is to advance global efforts on the threat of climate change by limiting global temperature rise during the 21st century to well below 2°C above pre-industrial levels. The agreement also has the objective of strengthening the capacity of nations to deal with the impacts of climate change. The agreement, which entered into force in November 2016, stands out for its virtually universal coverage across the international community of nations and the ambition of its objectives. Prior to the conference, participating nations were required to prepare and publish their Intended Nationally Determined Contributions (INDCs) as part of a new mechanism to allow each nation to be part of the climate solution to the best of its capabilities, with a focus on GHG emissions reductions within 2025–2030 and adapting or reducing vulnerability to climate change. The agreement established two processes, both on five-year cycles. The first process was a 'global stocktake' to assess global progress toward meeting the long-term goals of the agreement. The second was the submission of new NDCs by nations, a process which would be informed by the global stocktake. COP21 was a pivotal moment in the history of multilateral climate action and is especially significant because it is a legally binding agreement.

COP Conference	Highlights
COP22 Marrakech, 2016	The speed with which the Paris Agreement entered into force was testament to the political will at COP21. COP22 therefore had the task to determine the rules for the implementation of the agreement on matters such as how to measure progress and progressively grow ambition, how to raise finance for adaptation and mitigation, and how to account for international emissions trading. The election of Donald Trump as the new US president gave parties some cause for concern about the US contribution to future involvement. However, parties remained focused on implementation tasks of the agreement and set 2018 as their deadline for completing the necessary details to implement the agreement.
COP23 Bonn, 2017	COP23 was an interesting test of the resilience of the Paris Agreement, especially under the spotlight of the US announcement to withdraw and a renewal of old hostilities along developed versus developing economy lines. To be considered a success, the conference had to negotiate a number of issues. It had to show that the Paris Agreement would not live or die by US participation. It also had to create the foundation for adopting common rules on the implementation of the Paris Agreement. It also had the unenviable task of promoting the mainstreaming of climate objectives within the multilateral development agenda. Arguably, it went a long way to achieving those objectives. The conference did not seek milestone achievements; rather, it set out to pave the way for COP24 the following year, when parties would be expected to adopt the Paris rulebook and also conduct a major stocktake of collective progress.

COP Conference	Highlights
COP24 Katowice, 2018	COP24 was tasked with finalizing the Paris Rulebook, and especially focus on setting the framework for nations to measure, report, and verify emissions reductions as part of their NDCs. The conference discussed key issues of the Paris Agreement on mitigation, adaptation, market mechanisms finance, international cooperation, transparency, and the global stocktake. The conference was notable for its adoption of the Paris Rulebook to govern the implementation of the Paris Agreement. This arrangement is considered a robust mechanism for the communication of plans and the progress of NDCs. However, the conference was unable to hide cracks along traditional fault lines: developed economies were reluctant to commit to further emissions reductions or support for developing economies (especially significant in light of the recently released IPCCC special report on the impact of global warming of 1.5°C above pre-industrial levels), while developing economies were disappointed at the lack of focus on equity issues in the global stocktake.

COP Conference	Highlights
COP25 Madrid, 2019	COP25 highlighted the difficulties facing the international community to negotiate agreeable outcomes and address contentious issues left from previous COPs. It had an important role to play in not only bringing the 2015 Paris Agreement into force but also setting parties for more ambitious future emissions reductions commitments. Its objectives therefore included the development of guidelines on international carbon markets (Article 6 of the Paris Agreement), adaptation to climate impacts, loss and damage suffered by developing economies due to climate change, and climate finance for decarbonization. COP25 failed to adopt rules for international carbon trading, which was the last remaining piece of the Paris Rulebook to implement the Paris Agreement. At the same time, developing economies continued to complain about the unavailability of pledged resources to help them combat growing impacts of climate change. In the end, the conference succumbed to the typical developed versus developing issues, highlighting an increasing chasm of necessary action and slow responses from most major economies, so that, for instance, the Europeans found themselves at loggerheads with countries as diverse as the US, Russia, and Brazil.

COP Conference	Highlights
COP26 **Glasgow, 2021**	The Glasgow conference was delayed a year by the COVID-19 pandemic. It was notable as a summit that was characterized by a war of words. Fractions grew over the final text of an agreement on how to deliver on 'ambition' and the phasing out or phasing down of coal and fossil fuel subsidies, and of course aid for developing economies. Ultimately, compromise wording was settled on for these issues, but as with any compromise, there were many unsatisfied parties at the final outcome. There were some notable successes: COP26 delivered a Paris Agreement Rulebook for international cooperation through carbon markets – nearly six years after negotiations had begun on the matter. The rulebook allows countries to focus on their emission reduction targets. This was highlighted by more ambitious reductions commitments announced by countries like Brazil, Saudi Arabia, India, and Australia. COP26 also established a new International Sustainability Standards Board (ISSB) to harmonize global sustainability reporting practices. The Global Methane Pledge, signed by over 100 countries, was also announced, shining a spotlight on the growing opportunity for methane emissions reductions. There was plenty of frustration too, especially on the growing failure to realize the climate finance target of US$100 billion per year for developing economies and last-minute drama over wording to 'phase down' rather than 'phase out' unabated coal.

COP Conference	Highlights
COP27 **Sharm el-Sheikh, 2022**	COP27 was first conference held after the Paris Agreement Rulebook was agreed in Glasgow the previous year. There was also some strong geopolitical momentum from the US recommitment to the Paris Agreement through the Biden administration, as well as similar assurance from Brazil through the incoming Lula administration. The biggest achievement of COP27 was the establishment of the Loss and Damage Fund to help the most vulnerable countries cope with the devastating impacts of climate change. After over 20 years of marginal progress on the issue, a deal was finally hammered out that committed developing economies to funding solutions to the 'loss and damage' suffered by communities and the natural environment through climate change. Mitigation ambition remained unchanged from the Glaswegian event the previous year, and so unabated coal power was referenced with 'phase down,' and not expanded to include oil and gas. On adaptation, the final agreement text merely urged countries to urgently scale up climate finance toward adaptation solutions. The term 'nature-based solutions' was also referenced for the first time in a COP with regard to mitigation and adaptation actions.

COP Conference	Highlights
COP28 **Dubai, 2023**	Negotiations at COP28 concluded with the 'UAE Consensus.' COP28 was highlighted by the conclusion of the first global stocktake (GST) of the Paris Agreement. A notable element of the UAE Consensus was the agreement on a global transition away from fossil fuels in energy systems 'in a just, orderly, and equitable manner' – this was the first time that fossil fuels were mentioned in a COP (previously, coal was the sole mention). Similar to other COP outcomes, the agreement is a nonbinding one. And COP28 settled on an agreement to implement the Loss and Damage Fund. Methane also emerged as a big agenda item, with major economies, including the US and EU, committing to address their domestic methane emissions, and the World Bank launching its own methane program. COP28 also tacitly acknowledged the inadequacy of climate action alone, with significant delegate attention given to nature-based solutions and funding for nature and conservation initiatives. The Dubai conference also proved that adaptation had now become a mainstream focus area, with a large number of actors beginning to act on relevant socio-economic infrastructure, such as health, water, and food. Ultimately, COP28 demonstrated that momentum toward 'net zero' is building, but it also confirmed that significant hurdles remain in the quest for meaningful climate action.

COP Conference	Highlights
COP29 Baku, Azerbaijan, 2024	The 29th edition of the UN Climate Change Conference (the COP 29 summit) was held in November 2024 in Baku, Azerbaijan. The conference, dubbed 'the finance COP', was marked by a sense of progress and setbacks amidst a series of significant administrative and ideological challenges.
	The summit extended two days beyond its deadline. Protracted negotiations involving policymakers and leaders eventually delivered a climate finance deal, revealing the complexities and fault lines of multilateral climate diplomacy.
	One of the primary goals of the summit was to establish a new annual target for global climate finance. After intense discussions, participating parties agreed on annual US$300 billion disbursements to developing economies by 2035. However, this amount was met with dissatisfaction from many in the Global South, who considered it inadequate given the identified needs for climate adaptation and mitigation. Additionally, the long timeline extending to 2035 was criticized for potentially delaying necessary actions to facilitate the global transition to renewable energy sources. The improbable success of reaching a deal in the first place should not be sneered at, particularly in the face of increased socio-economic pressures from constituents of Western countries and the growing isolationist attitudes of these governments. Having a deal was in itself a remarkable achievement and may be a catalyst for future initiatives.
	Additionally, COP29 succeeded in establishing a framework for carbon credits, a decade in the making, allowing countries to establish credits to offset emissions either domestically or trade in a market. While the details regarding the structure of registries and transparency measures are yet to be fully defined, this resolution could potentially result in significant funding and support for projects aimed at reducing carbon emissions.

COP Conference	Highlights
COP29 (cont.)	The sense of accomplishment was counterbalanced by a few factors. For one thing, the election of Donald Trump as US President, an avowed climate sceptic, dampened the mood and contributed to a lowering of ambitions from expected reductions in US involvement in global climate efforts. With Trump pledging to withdraw the US from international climate efforts and appointing a climate sceptic as his energy secretary, optimism about future US support and leadership in climate initiatives had been significantly eroded. This was a significant barrier to the setting and achieving of ambitious goals at the summit. The spectre of the perennial north–south ideological fault lines appeared over the issue of financial responsibilities; most notably in the push by Western parties to include some developing economies as contributors to the US$300 billion target. After the COP29 presidency announced the deal, strong objections were raised on technical grounds. In fact, representatives of the Least Developed Countries (LDC) and Alliance of Small Island States (AOSIS) negotiating blocs stormed out on the last day, citing an obscure negotiation process that systematically shut them out of decisions. The current US$100 billion arrangement includes a component from multilateral development banks like the World Bank or IMF, which is counted as part of climate finance since these institutions contribute virtually half of their lending to climate projects. Developed countries wanted to count all of the World Bank or IMF shareholders in the US$300 billion goal, thereby implicitly including shareholding countries like China and India as contributors towards the US$300 billion. This technical alleviation of responsibilities for the Global North, frustrated the Global South and India was particularly vocal about this issue, arguing that developed, wealthier nations should bear a greater fiscal responsibility due to their historical emissions and economic capacity.

COP Conference	Highlights
COP29 (cont.)	As negotiations proceeded, many parties openly pondered the fragility of the multilateral process and issues related to the current trajectory of global climate efforts in the face of the growing intensity and frequency of extreme weather events (like flooding, heat waves and droughts). The impacts of these events underscore the need for global collaborations to facilitate the appropriate mitigation and adaptation responses.

VII

Future Views

"Good evening, sir. As you may know, the soaring costs of recent environmental-protection legislation have forced us to pass part of this burden along to the consumer. Your share comes to $171,947.65."

The influence of climate issues on world events

Climate change is an unprecedented global challenge that is reshaping the course of history and affecting various dimensions of human life. Its far-reaching implications touch every continent and influence a plethora of world events. This chapter delves into the multifaceted impact of climate change with a lens into the future.

North and South: Will climate change bridge the gap?

The potential impact of climate change on bridging the gap between the North and the South is a complex and multifaceted issue, deeply intertwined with geographical factors. These factors can significantly influence the extent to which climate change might contribute to narrowing or widening the development gap between these two regions. What are the key factors that will shape the interaction between climate change and global disparities?

Geographical factors, including proximity to coastlines, altitude, latitude, and resource availability, contribute to the differential vulnerability of regions to the impacts of climate change. Developing countries in the Global South, often characterized by their geographical positioning, are more susceptible to climate-related challenges such as extreme weather events, sea-level rise, and desertification. These vulnerabilities, rooted in geography, amplify the adverse effects of climate change, potentially exacerbating the gap between them and the more resilient developed countries in the North.

Resource availability is another significant geographical determinant that influences regions' capacities to adapt to climate change. Water scarcity, a pressing concern in various parts of the Global South, impacts agriculture and overall livelihoods. Geographical disparities in water resources can lead to differential vulnerabilities, with developed countries in the North being relatively better equipped to address water-related challenges. Their access to advanced technology and infrastructure enables them to mitigate the impacts of water scarcity more effectively than many countries in the Global South.

Moreover, the geographic diversity of agricultural landscapes plays a crucial role in shaping how climate change affects food production. Temperature, precipitation patterns, and soil quality influence agricultural productivity. As climate change leads to shifts in these factors, the North–South gap in agricultural output might widen. Developed countries in the North can invest in innovative, climate-resilient agricultural practices and technologies to maintain or even enhance food security. In contrast, many developing countries in the South, which heavily rely on agriculture for their economies and food supply, might experience reduced yields due to climate change, thereby exacerbating existing challenges in achieving food security.

One of the most significant geographical ramifications of climate change is the potential for climate-induced migration and displacement. Rising sea levels, intensified weather events, and changing agricultural conditions can force populations to migrate, often from vulnerable areas in the Global South to more developed countries in the North. While this migration could provide temporary relief to affected populations, it can also strain resources and potentially exacerbate social tensions in destination countries. The geographic redistribution of populations due to climate change could have broader geopolitical and social implications, thereby affecting the dynamics of bridging the North–South gap.

Biodiversity and ecosystem services, intricately linked to geography, are critical determinants of regions' abilities to adapt to climate change impacts. Many developing countries in the Global South are home to rich biodiversity and ecosystems that provide essential services such as water purification, carbon sequestration, and natural disaster mitigation. Climate change threatens these services, which are vital for both local livelihoods and global environmental stability. As these services erode due to climate change, the Global South's development challenges could intensify, potentially deepening the development gap between the North and the South.

Geographical factors also shape international collaborations and partnerships for climate action. Coastal and low-lying developing countries in the Global South are on the front lines of climate change impacts, particularly sea-level rise. These countries often seek assistance and cooperation from developed countries in the North to address these challenges. The geographical realities of vulnerability and shared risks can foster international cooperation that facilitates technology transfer, capacity building, and resource sharing. Such collaborations have the potential to contribute to bridging the gap by addressing the immediate and long-term impacts of climate change.

An intriguing geographical aspect is the Arctic region, where climate change is occurring at an accelerated rate. While sparsely populated, the Arctic's melting ice and changing climate have global implications. The effects of Arctic warming can influence weather patterns and contribute to sea-level rise, which affects regions both in the North and the South. This interconnectedness underscores the far-reaching geographical reach of climate impacts, highlighting the intricate relationship between various parts of the world.

It is clear that geographical factors play a pivotal role in shaping the interaction between climate change and the North–South development gap. Vulnerability, resource availability, agricultural productivity, migration patterns, ecosystem services, and international collaborations are all intricately linked to geographical features. While the impacts of climate change have the potential to either exacerbate or mitigate disparities between the North and the South, addressing these disparities requires a comprehensive approach that accounts for the geographic complexities of vulnerability, adaptation, and cooperation. As we navigate the challenges of climate change, acknowledging and addressing these geographical considerations is essential for fostering sustainable development and equitable outcomes for all regions of the world.

Shifting climate battles: The next wave of climate issues

Climate change has long been recognized as one of the most pressing global challenges of our time, with its far-reaching impacts on ecosystems, economies, and societies. The discourse surrounding climate change has primarily focused on the consequences of increasing global temperatures, rising sea levels, and extreme weather events. However, there is a growing realization that the ongoing shifts in the climate system will give rise to a new wave of challenges, which we can term 'shifting climate battles.' These battles are characterized by the intricate interplay between changing environmental conditions, resource availability, and human responses. Below we highlight five key areas where the emerging dimensions of shifting climate battles will play out, exploring their scientific underpinnings, potential implications, and the strategies required to navigate this complex terrain.

1. SHIFTING CLIMATE ZONES AND BIODIVERSITY LOSS

As the climate continues to change, ecosystems across the globe are experiencing shifts in temperature and precipitation patterns. One of the key challenges arising from this is the migration of climate zones. Species that are adapted to specific climatic conditions may face difficulties in adapting to new conditions or migrating to suitable habitats. Scientific models predict that these shifts could lead to substantial losses in biodiversity, as species unable to move quickly enough may face extinction.

For instance, a study by Urban et al. (2016) projected that up to one in six species could face extinction due to climate change by the end of the century. These losses can have cascading effects on ecosystem services, such as pollination, nutrient cycling, and carbon sequestration, further exacerbating the challenges posed by shifting climate zones.

2. AGRICULTURAL CHALLENGES AND FOOD SECURITY

Changes in temperature and precipitation patterns can significantly impact global agricultural systems, leading to reduced crop yields and altered food production patterns. The 'breadbasket' regions, which are crucial for global food production, may face increased risks of droughts, heatwaves, and changing pest dynamics. According to the Intergovernmental Panel on Climate Change (IPCC), by 2050 yields of major crops like rice and wheat could decline by up to 25% due to climate change.

The implications for food security are substantial. A study by Waha et al. (2017) estimated that climate change could increase the number of undernourished children by 20 million by 2050. Shifting climate battles in the agricultural sector necessitate the development of adaptive strategies, such as the breeding of climate-resilient crops and the adoption of precision agriculture techniques.

3. WATER SCARCITY AND CONFLICT

Alterations in climate patterns have profound implications for water availability, particularly in regions that heavily rely on seasonal precipitation or glacial meltwater. As the availability of freshwater resources becomes increasingly uncertain, the potential for conflicts over water resources rises. The Middle East and North Africa region, already prone to water stress, is particularly vulnerable to this phenomenon.

The connection between climate change, water scarcity, and conflict has been extensively studied. For instance, a report by the World Bank highlights that a 4°C warming scenario could lead to significant water scarcity in many regions, potentially triggering conflicts and displacement. The shifting climate battle over water resources underscores the urgency of effective water management strategies and international cooperation to prevent conflicts.

4. RISING SEA LEVELS AND COASTAL VULNERABILITY

Sea-level rise is a well-documented consequence of climate change, driven primarily by the melting of glaciers and the thermal expansion of seawater. Coastal communities, especially in low-lying areas, are facing the brunt of this phenomenon. Rising sea levels exacerbate the risks associated with storm surges and coastal erosion, increasing the vulnerability of infrastructure, ecosystems, and human settlements.

According to the IPCC's Special Report on the Ocean and Cryosphere in a Changing Climate (2019), under a high-emission scenario, global mean sea levels could rise by up to 1.1 meters by the end of the century. This projection underscores the urgency of adapting coastal regions through measures like building resilient infrastructure, restoring natural coastal buffers, and managed retreat from high-risk areas.

5. CLIMATE-INDUCED MIGRATION AND DISPLACEMENT

The convergence of climate change impacts, such as extreme weather events, sea-level rise, and resource scarcity, can contribute to the displacement of populations. This form of climate-induced migration poses complex challenges, ranging from the provision of humanitarian assistance to the integration of displaced communities into host societies. It also highlights the ethical and legal dimensions of responsibility and assistance toward climate refugees.

The Internal Displacement Monitoring Centre estimates that, in 2020, over 30 million people were newly displaced within their own countries due to natural disasters, many of which were linked to climate change. This statistic underscores the need for comprehensive international frameworks that address the rights and well-being of climate migrants.

Shifting climate battles represent the next phase of challenges in the context of climate change. These battles emerge from the intricate interactions between changing environmental conditions and human responses, giving rise to consequences that span ecosystems, economies, and societies. Scientific evidence underscores the urgency of addressing these challenges through proactive measures that encompass biodiversity conservation, resilient agriculture, water resource management, coastal adaptation, and inclusive policies for climate migrants. The shifting climate battles call for global cooperation, innovation, and adaptive strategies to navigate the complex terrain of a changing climate. As we continue to grapple with the impacts of climate change, it is imperative that we recognize and respond to these emerging battles to secure a sustainable future for generations to come.

2101: What to expect?

TEMPERATURE TRAJECTORIES AND THEIR CONSEQUENCES

Projected climate scenarios for the year 2101 offer insights into potential global temperature trajectories based on different levels of greenhouse gas emissions. Under the business-as-usual scenario, where emissions continue unabated, the global average temperature could surge by

3.7–4.8°C by the end of the century. This trajectory would amplify the frequency and intensity of extreme weather events, disrupt ecosystems, and intensify sea-level rise.

Conversely, adopting ambitious mitigation strategies in line with the Paris Agreement's 1.5°C target offers a more optimistic outlook. By drastically reducing emissions, the global temperature increase could be limited to 1.5–2.0°C. While this represents a considerable challenge, it would yield substantial benefits, including the preservation of vital ecosystems, reduced risks of irreversible impacts, and enhanced societal resilience.

Balancing these scenarios requires a comprehensive shift toward low carbon energy sources, sustainable land use, and carbon capture technologies. It underscores the need for international collaboration to ensure that the most vulnerable regions receive adequate support in transitioning toward a low-carbon trajectory.

OCEANIC CHANGES AND RISING SEA LEVELS

The projected climate scenarios for 2101 also point to significant changes in the world's oceans, driven by warming temperatures and melting ice sheets. Sea-level rise is an immediate concern, with potential consequences for coastal communities and critical infrastructure. The melting of polar ice sheets, including those in Antarctica and Greenland, could contribute to substantial sea-level rise over the long term, with estimates ranging from a few centimeters to several meters by 2101.

Rising sea levels would intensify the frequency and severity of coastal flooding and storm surges, increasing risks for millions of people living in vulnerable areas. It would also exacerbate saltwater intrusion into freshwater sources, impacting agriculture and freshwater availability. Adaptation strategies, such as constructing resilient coastal infrastructure and implementing sustainable land use practices, are essential to minimize the potential impacts of rising sea levels.

ECOSYSTEM RESPONSES AND BIODIVERSITY SHIFTS

Climate change projections for 2101 indicate that ecosystems across the globe will undergo transformative shifts in response to changing climatic conditions. These shifts will manifest as changes in species distribution, altered migration patterns, and potential extinctions. Coral reefs, critical marine biodiversity hotspots, are particularly vulnerable to temperature increases and ocean acidification. Widespread coral bleaching events could threaten the delicate balance of these ecosystems and have cascading effects on the marine life and coastal communities that rely on them.

Terrestrial ecosystems are not immune to these changes. Species will need to adapt to new environmental conditions, and some may struggle to survive in altered habitats. The shifting ranges of plant and animal species could disrupt ecosystem services such as pollination, water purification, and carbon sequestration. Conservation efforts must prioritize the protection of biodiversity-rich regions, implementation of habitat corridors, and the preservation of critical ecosystems like forests and wetlands.

IMPACTS ON SOCIETAL AND ECONOMIC SYSTEMS

The projected climate scenarios for 2101 bear profound implications for societies and economies around the world. Increasing temperatures, along with changes in precipitation patterns, could lead to decreased agricultural yields and food insecurity, affecting vulnerable populations the most. Heatwaves and extreme weather events could also pose health risks, particularly to the elderly and those with pre-existing conditions.

Coastal communities face heightened risks due to sea-level rise and storm surges. The potential for increased displacement and economic losses underscores the need for comprehensive adaptation strategies, including urban planning, early warning systems, and resilient infrastructure. Additionally, the economic costs of climate change-related damages could amount to trillions of dollars annually by 2101, emphasizing the urgency of mitigative actions.

TECHNOLOGICAL AND POLICY SOLUTIONS

Navigating the challenges posed by climate change in 2101 requires the rapid deployment of innovative technologies and robust policy frameworks. Carbon capture and storage (CCS) technologies hold promise for mitigating emissions from industries that are hard to decarbonize. Advancements in renewable energy sources, such as solar and wind, can play a pivotal role in reducing dependence on fossil fuels.

Comprehensive climate policies are paramount, with carbon pricing mechanisms incentivizing emission reductions and the preservation of carbon sinks like forests. Policies focusing on sustainable land use and deforestation prevention are critical in maintaining healthy ecosystems that act as carbon reservoirs. The development and dissemination of these technologies and policies require international cooperation and sustained commitment from governments, businesses, and civil society.

The path forward: navigating uncertainties

At the midpoint of the 2020s and as 2101 approaches, accompanied by scorching temperatures, international conflict, economic pressures, and devastating bushfire seasons, the world has focused growing attention on annual global climate conferences. Against this backdrop, the international community has continued its deliberations over climate change to increasing attendances. These deliberations reflect the issues discussed in earlier chapters.

In 2023 the world turned its eyes on the UAE which hosted the COP28 conference. As a previous chapter has already discussed, the outcomes of climate negotiations are largely a function of the formation and dynamics of groupings. As these groupings have formed into formal and informal blocs, group dynamics and international geopolitics have become as much, if not more, of a determining factor of the outcomes as scientific discourse.

During negotiations, blocs give support to specific policies or viewpoints and consequently climate negotiations have become an extremely political process and one which must delicately balance scientific and political views. It is very difficult to get the science and the politics into perfect alignment but this is precisely what climate negotiators are often attempting to achieve. Each bloc tends to have issues that it strongly advocates for; for instance, the Umbrella Group made up of developed and high-emitting economies like the USA,

Australia, and Canada, tends to push for the reduction of GHG emissions from all countries and not just industrialized economies. On the other hand, the bloc of Least Developed Countries (LDCs) tends to focus on capacity building as an integral component of climate solutions. The LDCs and the Umbrella Group are in philosophical opposition to one another on the idea of current or historical responsibility for GHG emissions reductions. It is these kinds of differing viewpoints that have shaped climate policy outcomes since the advent of the United Nations climate system.

By the time COP28 kicked off, a number of key issues had surfaced to the top of the global climate policy agenda after many years of discussions. The nature and outcomes of discussions on these issues will be pivotal to the direction of global climate solutions over the coming decades. They will also have repercussions for international relationships and the global economy.

Climate change has become one of the most pressing issues of our time. Shifting climate patterns and extreme impacts are already being felt across the world, and the critical need for strong action on climate change has gained significant traction in recent years. The Paris Agreement has become a symbol of the need for robust climate action, with its overarching goal to limit global warming to well below 2°C becoming a rallying call across industry, politics, and various social fields. This goal will only be achieved through significant efforts in innovation, policy-making, and financial investment in green technologies to slash human-induced greenhouse gas emissions.

But there are other non-greenhouse issues that now also represent a bigger part of the problem. Along with climate change, biodiversity loss poses another significant threat to global sustainability. Conservation efforts have brought attention to rapid loss and fragmentation of species diversity as a result of human activities such as habitat destruction, pollution, overexploitation, and the effects of climate change. Biodiversity is essential for the healthy existence of living beings on the planet. Without a wide range ("diversity") of animals, plants and microorganisms ("biological"), it would not be possible to have healthy ecosystems to sustain the life of all species, including human beings. The interconnectedness and relationships between different species are such that all species rely on numerous other species for life. Biodiversity is also intrinsically valuable in its own right, as each species has a right to existence and life on the planet. Finally, biodiversity "capital" is important to human existence as it provides services such as pollination, water purification, and carbon sequestration.

As climate change and biodiversity have grown immensely in profile over the last few years, the discourse on social equity and justice has developed relatively modestly over the same time. However, it is no less important; indeed, the United Nations Sustainable Development Goals are underpinned by the principle of inclusiveness of solutions and leaving nobody behind. Climate solutions will only be truly meaningful if they can address existing social inequalities, promote access to education and healthcare, and empower marginalized communities. This is the only way that wider sustainable development benefits can be shared in a just and equitable manner, thereby ensuring the "sustain" in sustainable development.

Technological innovation has also become a significant pillar in the field of climate change and sustainability at large. Solutions have developed at a rapid rate in areas such as sustainable materials and they are changing our interactions with the natural environment. With the growing influence of digital technologies and the Fourth Industrial Revolution, the scope for monitoring and managing environmental impacts and optimizing resource use is growing wider and wider. However, in the name of social equity and justice it is also important that these technologies are accessible across all communities and nations and that their deployment does not reinforce existing inequalities or promote environmental damage.

The need for international collaboration and governance has been identified as a key enabler of mobilizing necessary resources and institutional capacity to drive climate and sustainability solutions. Significant levels of coordinated multilateral responses are needed for global challenges of such a scale. The United Nations is currently the most comprehensive system for driving such efforts through its network of international frameworks, agreements and bodies to facilitate international cooperation. The UN system is a also a significant partner to foster education and public awareness across the world. This is particularly important because it has become apparent that an informed public can be particularly influential toward policymaking and the integration of sustainability into the formal aspects of education such as curricula and the informal aspects such as cultivating behavioral change.

Over the years, climate negotiations have invariably revolved around these issues and at the time of publishing this book, the most significant recent climate conference was the 2024 UN Climate Conference of the Parties – COP29 – hosted by Azerbaijan at the end of 2024.

COP30 and subsequent conferences will take on the baton of issues and aim for resolution on the key points.

Looking ahead

As one looks into a figurative crystal ball, it is clear that issues of discussion in the years ahead will fall into two broad categories: technical and administrative issues on the one hand and ideological issues on the other.

TECHNICAL AND ADMINISTRATIVE ISSUES

As COP29 concluded, a number of technical and administrative issues remain part of an ongoing and growing inbox of requirements. These include:

Nationally Determined Contributions (NDCs): Parties to the Paris Agreement were required to provide their new NDCs by February 2025 in preparation for their official consideration before COP30 in November 2025. There were some doubts over the extent to which parties would be able to meet the disclosure deadline, which in turn would compromise the ability to implement outcomes from the first Global Stock Take.

Climate finance: Under the Paris Agreement, parties agreed to a new climate finance target by 2025 to direct funds towards developing economies to tackle climate change. But this 'new collective quantified goal' (NCQG) has been fraught with significant issues at every turn, including disagreements on amount of funds, type of funding, target of funding and related matters. The developing economies of the Global South are likely to increase pressure for greater clarity on how developed economies of the Global North intend to increase the annual finance goal (beyond US$100 billion).

Multilateral financing: There is growing pressure for Multilateral Development Banks (MDBs) to integrate global challenges like climate change into their agendas. It appears that MDBs in general have a long way to go to implement such required reforms. They hold trillions of dollars in assets globally and disburse funds for climate mitigation and adaptation across the world; they have significant capacity to fund climate change projects but there is pressure for them to take on more risk to expand their lending.

IDEOLOGICAL ISSUES

Ideological differences have existed in a spectrum from nuisance to severe stalemate at global climate negotiations almost since day one. These differences tend to revolve around the attribution of responsibility and capacity to address climate change:

North and south: In essence, climate negotiations have always been shaped by equity issues between the developed economies of the Global North and developing economies of the Global South, particularly over responsibility for GHG emissions and cost burdens related to mitigation and adaptation. These differences show no sign of 'abating' and may even intensify over climate finance modalities and phasing discussions.

Trade tensions: Future years will likely witness more contentious discussions on climate-related trade policies. Some developing economies are already expressing concerns over the possible impacts (mostly through climate costs) of climate-related trade policies such as carbon border levies. Equally, it is possible that 'retaliatory' climate-related trade policies may be initiated by either developing or developed economies as states continue jockeying for favourable climate-related trade positions.

Useful books

Governing Climate Change
Harriet Bulkeley, Peter Newell
2015
Governing Climate Change, Second Edition, provides a short and accessible introduction to how climate change is governed by an increasingly diverse range of actors, from civil society and market actors to multilateral development banks, donors, and cities.

The Politics of Climate Change Knowledge
Nowrin Tabassum
2022
This book addresses political knowledge of climate change and its relation to labelling people affected by climate change, either as 'climate refugees' or as 'climate change-induced displaced people or migrants.'

Climate Change and Sustainable Development: Prospects for Developing Countries
Anil Markandya, Kirsten Halsnaes
2021
This text argues that the policies pursued by developing countries will be crucial in determining the progress of climate change. Many are industrializing rapidly and the largest, particularly China and India, could have an impact at least as significant as that of the already industrialized economies – the reason given by President George W. Bush for taking the US out of the Kyoto Protocol. The future of sustainable development in large measure depends on developing countries.

www.ingramcontent.com/pod-product-compliance
Lightning Source LLC
Chambersburg PA
CBHW051258020426

42333CB00026B/3257